Praise for
NUMBERCRUNCH

'The perfect introduction to the power of mathematics –
fluent, friendly and practical.'
Tim Harford, author of *How to Make the World Add Up*

'Numbers don't lie but they often speak a foreign language.
Professor Oliver Johnson is a superb maths whisperer on
a mission to arm his readers with the tools to distinguish
sound claims from the many phoney ones that bombard us
every day. An invaluable addition to the
modern baloney detection kit.'
Ananyo Bhattacharya, author of *The Man from the Future*

'A clear, straightforward, informative guide to
understanding numbers. I wish I'd read it years ago.'
Tom Chivers, author of *How to Read Numbers*

'Oliver Johnson provides useful and timely insights by
deploying simple but powerful mathematical techniques.
Numbercrunch shows how to apply crucial ideas to a range of
real-life problems. A valuable and topical guide to navigating
the world of numbers, from sports transfers to medical testing.'
Adam Kucharski, author of *The Rules of Contagion*

D1431297

NUMBERCRUNCH

Oliver Johnson is Professor of Information Theory and Director of the Institute for Statistical Science in the School of Mathematics at the University of Bristol. He was previously a research fellow at the University of Cambridge and Fellow of Christ's College, Cambridge. He has frequently appeared on BBC Radio 4 and written for the *Spectator*, and has been quoted in a variety of newspapers including *The Times*, *Guardian*, *Daily Telegraph* and *New York Times*.

Oliver is on Twitter as @BristOliver, where he tweets about maths, music and Aston Villa. He lives in Bristol.

NUMBERCRUNCH

A MATHEMATICIAN'S TOOLKIT FOR MAKING SENSE OF YOUR WORLD

Professor Oliver Johnson

First published in the UK by Heligo Books
An imprint of Bonnier Books UK
4th Floor, Victoria House, Bloomsbury Square,
London, WC1B 4DA

Owned by Bonnier Books
Sveavägen 56, Stockholm, Sweden

Hardback – 978-1-788708-33-3
Trade paperback – 978-1-788708-34-0
Ebook – 978-1-788708-35-7
Audio – 978-1-788708-36-4

A CIP catalogue of this book is available from the British Library.

Designed by EnvyDesignLtd
Printed and bound by Clays Ltd, Elcograf S.p.A

1 3 5 7 9 10 8 6 4 2

Heligo Books is an imprint of Bonnier Books UK
www.bonnierbooks.co.uk

To all my maths teachers
and lecturers

'The essence of mathematics is not to make simple
things complicated, but to make complicated things simple.'
Stanley P. Gudder, *A Mathematical Journey* (1976)

Contents

1
Introduction

Over the years, I have come to expect certain reactions when I tell people that I'm a mathematician. Often, it's a nervous edging away and a sudden need to be at the other end of the room. Usually people will announce, 'I was never any good at maths at school,' in a way that implies they are almost proud of this fact.

However, recently this has started to change. Alan Turing is on the £50 banknote. Movies like *Moneyball*, *Hidden Figures* and *Good Will Hunting* have been nominated for Oscars. The Breakthrough Prize ceremony sees Hollywood stars giving awards worth millions of dollars to mathematicians in a televised ceremony. Dare I say it, the subject might be becoming cool.

Another important factor in the rising profile of maths has been the coronavirus pandemic. Suddenly, numbers were everywhere. The latest data visualisations were being shared on social media like Kardashian Instagram posts. Words like 'exponential' and 'confidence interval' were being bandied around. At last, I have a sense that understanding numbers, being able to spot trends and make predictions are nothing to be ashamed of.

During the pandemic I tweeted a mathematical perspective on COVID statistics as @BristOliver, doing my best to help people understand what was going on and to make sense of the barrage of numbers. Doing this constantly reminded me that the maths skills that I teach my undergraduate students at the University of Bristol are incredibly valuable for all of us.

Our everyday lives are increasingly ruled by data and by algorithms. We can make ourselves understood by talking to Siri, and instantaneously receive almost professional-quality translations of foreign language text using Google Translate. Netflix can match our previous viewing to profiles of similar users, to recommend the next box set that we are most likely to get hooked on.

However, it may be less obvious that this kind of 'artificial intelligence' or 'machine learning' emerged from mathematics and from statistics. These ideas have had a rebrand for the 21st century and been supercharged by ever-increasing computing power, but it's always mathematics under the bonnet. These Silicon Valley marvels rely on ideas like the geometry of clouds of points in a world of millions of dimensions, techniques for finding structure and form in randomness and mathematically rigorous ways of dealing with vast amounts of data.

Most people don't know what a professional mathematician does all day. Perhaps they imagine that we are memorising harder and harder times tables ('one 19,573 is 19,573, two 19,573s are 39,146') or competing to see who can remember the most digits of pi. Maybe they imagine a dusty old man writing incomprehensible chalk equations full of Greek letters on a blackboard (and to be fair, this isn't always so far wrong). Some of this is the mathematicians' fault. We haven't exactly gone out of our way to explain why what we do matters.

This book is an attempt to redress that balance. I believe that navigating everyday life now requires making sense of endless numbers and facts, as more and more data about an ever-increasing number of fast-evolving global scenarios is available faster than ever before via our phones, tablets and computers. Further, I believe that mathematical ideas offer the right way of thinking about the world, to make sense of this barrage of data and of complex situations, and to avoid being led astray by misleading interpretations of them. I would like to teach you some of those tricks, and to help you understand the world through a mathematician's eyes. Twitter is a fantastic place to get short and timely messages across, but it doesn't allow you to show your working as mathematicians like to. I will try to do that here and explain why I think about things in the way that I do.

At the heart of all this lies the idea of a mathematical model. In a sense, we all have a mental model of how things work. For example, we know that if we drop something, then due to gravity it will fall to the ground. However, it was Isaac Newton's fundamental work in turning this idea into mathematical equations that meant that the effect could really be *understood*.

A model should have two features: it should help explain the data that we have and make predictions about situations that we have not been in yet, ideally with a well-calibrated degree of caution about the accuracy of these predictions. Mathematicians often work by spotting patterns in numbers or elsewhere, and by producing a theory that seems to explain them. Because of Newton's equations, even before Neil Armstrong stepped on to the moon from Apollo 11, we had a good idea of how strong the force of gravity would be there; and indeed, understanding these equations had allowed NASA to build a rocket that would arrive there in the first place.

A strange thing is that a good model does not need to be *exactly* correct. The statistician George Box famously wrote that 'all models are wrong, but some are useful'. We know from Albert Einstein's work on relativity that Newton's equations are not perfect, that for example they fail to address issues that arise from travelling close to the speed of light. However, in a 17th-century world of horse-drawn carriages and sailing ships, Newton's equations were accurate enough to make useful predictions about the behaviour of everyday objects. Even today, since even our fastest rockets don't approach light speed, from a practical point of view 'Isaac Newton is doing most of the driving right now', as Apollo 8 astronaut Bill Anders famously remarked. In the same way, while some mathematical models of the spread of coronavirus were undoubtedly too simplistic and ignored real-world effects, if they could make reliable short-to-medium-term predictions, that might well be enough for us.

Related to this is the idea of a 'toy model'. This can be a complete abstraction that may not bear much resemblance to the real world, but which nevertheless reveals something about its properties. There is a notorious and apocryphal story of a mathematician asked to consider the design of dairy farms, whose resulting model began with the words 'Consider a perfectly spherical frictionless cow in zero gravity . . .' However, well-chosen toy models can be extremely useful for thinking about the real world. I spent a certain amount of time thinking about how we would detect a coronavirus vaccine effect in data by imagining gradually giving out magic hats that gave instant protection from disease.[1] If we can abstract the key features into this kind of simple model, then the insights that we develop can be generalised back into everyday life, albeit with a certain amount of caution about the validity of the toy model.

There have been many excellent popularising books about mathematics. But they have often concentrated on the outré or exotic (Marcus Du Sautoy's *Finding Moonshine*, Simon Singh's *Fermat's Last Theorem*) or the relationship between mathematics and physics (Graham Farmelo's *The Universe Speaks in Numbers*).

Sometimes, in a popular understanding, the world of maths is reduced to one of solving puzzles and performing tricks with numbers (the mathematician and comedian Matt Parker excels at this). Maths can indeed be fun, and these kinds of games can be a fantastic way to attract people to the subject. However, I think it's also important to remember that maths is *important*, and drives much of the modern world. It offers a practical toolkit for making sense of things, and I would like to help you understand how to do that.

I'm not going to bombard you with equations and Greek letters. In fact, there aren't really going to be many equations at all. This is a book about maths as a way of thinking, not as a source of algebra problems. If I can, I'd rather draw the right picture to explain something than dive into serious calculations. I also know that people learn mathematics best when they do it, rather than just hear about it. For this reason, each chapter ends with suggestions of

---

1   This would produce a noticeable effect when plotting data on a logarithmic scale (see Chapter 3), where we would see a gradually steepening curve. Indeed, this effect could be seen in UK COVID death data in the spring of 2021. Although the assumptions of the model were undoubtedly too simplistic, it nonetheless gave a way of thinking about how to spot a vaccine effect.

ways to test your understanding and explore the themes that I have discussed. Nobody is going to be marking this homework, but I'd love to hear how you get on.

Most of the book naturally divides into three parts, each made up of four chapters. The first part deals with STRUCTURE. Mathematics is an excellent tool for understanding how the world works, partly because many important processes follow simple rules. Mathematics serves as a language and a formulation that captures this. It is no coincidence that the fundamental laws of physics are expressed using mathematical equations, for example.

However, often we don't directly observe the scientific laws governing the processes that we care about. We may only see a snapshot of this behaviour and want to infer what is going on by thinking about the way that these kinds of processes normally behave and the rules that drive them.

In Chapter 1, I will explain how plotting data on a graph is an excellent way to start doing this. However, this needs to be done carefully, because short-term patterns in the data can lead you into making overconfident predictions.

Chapter 2 takes a diversion into the world of numbers themselves. The modern world constantly presents us with a barrage of bewildering data, from the latest economic figures and scientific findings to opinion polls and sports results. I will give some tricks for understanding these numbers better, to approximate them and make sensible ballpark estimates of complicated quantities.

Chapter 3 will introduce you to the idea of exponential growth, whether that be in football transfers, bacteria or nuclear reactions, and will show how this can be represented on a graph using what mathematicians call a log scale. I will explain how doing this can give us insights into the behaviour of the coronavirus pandemic and of the stock market.

In Chapter 4 I will show that systems following simple rules can exhibit complex behaviour. By describing the motion of a pendulum and introducing an area of maths which arose from attempts to predict the weather, I will show how these kinds of mathematical models can capture the way that many processes operate.

Having described how mathematics can capture and describe structure, it might feel paradoxical to step away from this kind of neat and well-ordered world. However, it turns out that an often-contradictory impulse can also be described by similar tools. This is the tendency towards RANDOMNESS, which we will learn about in Part 2.

Our human instincts are often misled by the notion of randomness. For example, if the number 31 came up in the last two weeks' lottery draws, you might believe that this week it is less likely to come up (because things balance out on average) or more likely to come up (because this ball is clearly popular), whereas in fact the odds haven't changed at all. People are very good at fooling themselves that a pattern exists when really there is none. By understanding randomness, both in terms of expected behaviour and the extremes of what is likely, we will come to understand these kinds of issues better.

Chapter 5 introduces the key idea of randomness in human activities. By thinking about a simple coin toss, we will see what should happen on average if we repeat an experiment enough times, and how randomness and predictability can exist together.

Chapter 6 builds on this, by describing ideas from statistics such as confidence intervals, which mathematically express the underlying uncertainty about what is going on and allow us to take principled decisions as to whether to license a new drug, for example.

Chapter 7 explains how thinking about probability in the right way, taking account of information that we already have, can give insights into many important problems. For example, we will see how confident we really should have been in the results of coronavirus tests, and how to diagnose the causes of inequality in a company's hiring procedures.

Chapter 8 gives another perspective on probability, motivated by the idea of odds offered by bookmakers. We can move between probabilities and odds in a natural way, and we will see that bookmakers' odds are often a natural object to work with. This way of thinking helped Alan Turing and his colleagues at Bletchley Park to crack the Enigma code, but I will describe how it also gives us a better way to understand medical testing and can predict how market share of a new product or COVID variant will change over time.

Having understood randomness and structure, the third part of the book describes INFORMATION, a key final strand in understanding the modern world mathematically. As well as being the building block of much of everyday communication and media consumption, information and uncertainty can be quantified mathematically using a quantity known as entropy. We will see how information can slip into misinformation, can help us describe the evolution of the stock market and pandemics, and can even be competed over as a resource. Once again, all these things can be captured in a mathematical framework.

Introduction

In Chapter 9 I will tell you about one of my mathematical heroes: Claude Shannon. I will explain how his work tells us to think about the news sources that we consume, and how his ideas underpin the world of mobile phones and data downloads, as well as motivating an unexpectedly efficient way to test for disease.

Chapter 10 describes how a drunkard staggering home from the pub helps us to understand how rumours or computer viruses can spread across a network of connections. Again, this kind of behaviour is random and yet predictable, and understanding its properties moves us into a world of randomness which is much richer than the simple world of coin tosses.

Chapter 11 discusses further issues to do with randomness and noise. It shows how we can fool ourselves into believing that data contains patterns, and how other issues with the way that data is measured and reported can lead us to misleading conclusions. By understanding these issues, you will be better equipped to make sense of news stories containing complex data.

Chapter 12 describes what is known as game theory, which considers questions of collaboration and competition with one another. These kinds of mathematical ideas, which have applications in economics and biology, show us that often the right strategy in these sorts of situations involves a mixture of actions.

Finally, the book concludes in Chapter 13 with a plea for humility when thinking about complex situations. I discuss several ways in which we can fool ourselves into misreading a situation and give some suggestions for how we can avoid these errors.

Although it was first inspired by thinking about the coronavirus crisis, this isn't a book about the pandemic. There were already many excellent books about disease and infection, even before the media started talking about R numbers and herd immunity. Although some of my examples arose during the pandemic, there will also be examples from a much wider range of settings to show you how mathematicians and statisticians think.

I will describe a mathematical toolkit which comes in these three parts, which will enable you to understand the structural principles underlying the way the world changes, make sense of the randomness and uncertainty governing the way these things are reported, and to distinguish between information and misinformation.

Once you become aware of these ideas, you'll find more and more ways to use them to make sense of the world. It's almost impossible to predict what the dominant news story in ten years' time might be, but whatever it is, having this toolkit will put you in a better position to analyse it in a rational way and to distinguish between the signal and noise. If this book can help you develop a mathematical way of thinking, and to see the latest information through a mathematician's eyes, you will be ready to make sense of things for yourself.

PART 1:
STRUCTURE

Chapter 1
The right picture is worth a thousand words

Graphs and representations

Imagine you are a café owner who has started a promotional campaign for your new range of iced buns. You've advertised in the local paper, put posters on the lamp posts in your neighbourhood and left flyers in the post office. You would like to know if this has worked, and so will keep track of bun sales in successive weeks. If I simply tell you that these values were 143, 136, 147, 144, 149, 147 and 153, that stream of numbers can be hard to make sense of by eye. There is a danger of what comedians Mitchell and Webb referred to as 'Numberwang', as we are bombarded by a succession of figures without much context.

Indeed, the modern world is dominated by numbers. Whether it be the size of a national budget, the number of unemployed people or today's Bitcoin price, much of our daily discourse is driven by numerical values. This can perhaps seem intimidating. Numbers can sometimes be used to obfuscate an issue or can be presented without context.

I would like to help you make sense of numbers, to feel less daunted by them, and to understand the rules that govern their ebb and flow. This is something that can come with practice, and certain tricks can help us to understand numbers much better. The simplest of these is perhaps to draw a picture – or to be more precise, to plot them on a graph.

Graphs are a mixed blessing. Done well, they can be an extremely effective way to communicate. Done badly, they can be baffling. Further, because they can be so persuasive and are easy to share online without context, plots of data can even be a tool of misinformation. For this reason, it is important to understand exactly what a graph is telling us, and how to read it. To do this, we need to go back to mathematical fundamentals.

As an example, the majority of COVID graphs we were bombarded with by governments and health officials during the pandemic represented *time series* of data, in a way that is easy to make sense of. This kind of graphical representation of numbers can be extremely useful in other contexts too, although we need to be careful. For example, plotting your bun sale numbers apparently shows some structure.

Looking at the graph, the trend seems clear. There seems to have been a major increase in sales over these seven weeks. Generally, the numbers seem to be going up: the very low figure of 136 occurs close to the start, in week 2, and the highest figure of 153 occurs on the final week. I could imagine drawing a line on this graph that

The right picture is worth a thousand words

Bun sales per week

would be sloping up – the advertising campaign seems to be working! However, we need to be careful before jumping to conclusions.

It is easy to only look superficially at the picture, and not think about what it really represents. One thing to keep an eye on is the labels on such graphs. In this case, I've put the labels of the dates across the bottom (this is what we call the x-axis) and put the labels of the sales along the side (this is what we call the y-axis). You should notice that I have cut off the y-axis: the numbers don't run all the way down to zero, which has a serious effect. Here is the same data, plotted on a y-axis that starts at zero.

Suddenly, the same data looks quite different. There may still be an upwards trend, but it appears to be nothing like as dramatic as previously. It may simply be a case of fluctuations in the data; perhaps the advertising campaign is having no effect after all?[2]

---

2   To decide this question in a more principled way, we would need to think about randomness, in a way that I will describe in Part 2 of the book.

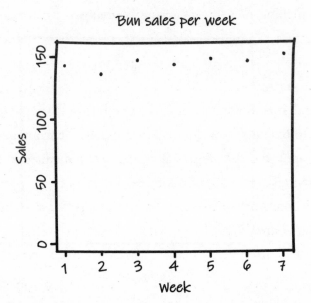

Bun sales per week

In general, cutting off the y-axis can make relatively minor changes seem more important. On many occasions, this kind of truncation is a perfectly legitimate choice to represent data. For example, when plotting daily temperature graphs for the UK, only the most absurd pedant would insist that the graph should start at −273 degrees Celsius because that represents the physics concept of absolute zero. However cold it may feel on a winter's morning, we will not experience those kinds of temperatures, and so having the graph start at that level would compress important daily variation into a very narrow range.

However, this sort of manipulation of the y-axis is a trick beloved of certain political parties producing bar charts for their election leaflets, and you should certainly watch out for it. It is important to remember that it is not the only way that data can be manipulated to give false impressions. Other tricks to look out for are cherry-picking of data, for example giving a comparison only between certain carefully selected items, presenting numbers over a carefully chosen period to present the desired conclusion, or even not labelling axes at all.[3]

In general, it is worth narrating the graph to yourself: what story is it trying to tell, and how does it achieve that? I encourage you to seek out graphs and data representations in the media and online, and to start looking at them in a more critical way – to become a more active consumer of these kinds of plots, rather than just passively letting your eyes glide over them.

There are deeper mathematical issues that are important to consider. To understand more, it is worth thinking about what we are trying to achieve with these data visualisations. The reason we are plotting data is to obtain insight into the process that generates it. Ultimately, we are looking for a mathematical model that helps explain these numbers, and that would allow us to predict future values and plan accordingly.

Functions, linear and polynomials

In its simplest form, this model might take the form of what mathematicians refer to as a *function*. This is simply a rule that works like a computer program: given a particular input, it will produce an output. In the iced bun example, the input would be the number of weeks that have passed, and the output would be the number of buns sold.

---

3 If you want to learn more about misleading ways that data can be presented, the book *Calling Bullshit: The Art of Skepticism in a Data-Driven World* by Carl Bergstrom and Jevin West comes highly recommended.

We can try to make sense of what kinds of functions we might expect to see. The absolute simplest case would be a function that didn't change at all. In other words, the output would be the same each time, which we often refer to as the *constant function*. Obviously, this isn't a very interesting example, but it is worth thinking about what you would see if you plotted it: the points would simply form a horizontal line.

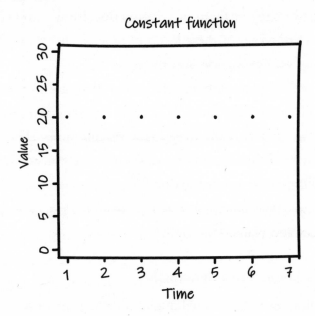

We could describe the next-simplest example in terms of an idealised version of the Voyager space probe heading into deep space under its own momentum, without power or friction. Isaac Newton's Laws of Motion tell us that such a probe would continue at a constant speed. If we measured at the same time each day, we would find that its distance away from Earth had increased by the same amount each time. If we plotted these daily values of distance on the graph, successive points would move upwards by the same amount, giving a straight line sloping upwards. Mathematicians call this a *linear function*.

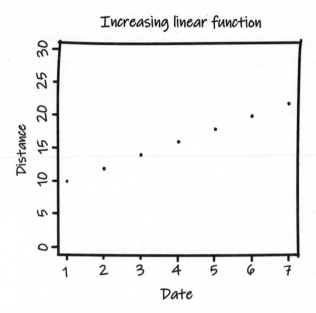

Although they are simple, linear functions are extremely useful to think about when understanding how things change.[4] There are two interesting things to notice. First, the steepness of the line carries *information*: the slope essentially tells us the speed of the probe – how far it has moved each day. The faster that Voyager is moving, the steeper the slope we observe. In fact, by plotting the daily positions and measuring the slope of the line we can calculate the rate of travel of the probe.

Second, it is interesting to think that the linear function emerges from a situation of 'no effect'. As described, the probe undergoes no change in speed, and left to itself would continue like that

---

4   Indeed, Newton's other great development, differential calculus, gives us a way to think of any function as being built up out of linear parts – but you'll be relieved to hear that I will not discuss that in any detail here.

indefinitely. As a result, the graph of the linear function would form the same straight line forever; linear functions are extremely predictable. We can comfortably extrapolate them by eye out into the future and see what values to expect at a later point.

Incidentally, there is no reason that linear functions must slope upwards, in the way that I have described here. From the point of view of an alien sitting on some distant planet with Voyager heading towards it, the distance to the probe would decrease by the same constant amount each time. Plotting these points on a graph, we would see a line sloping downwards.

We don't have to restrict our attention to processes that change

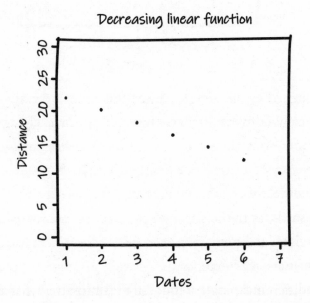

in a linear way like this. More interesting behaviour is possible. For example, suppose Voyager were not simply drifting in space, but continually firing a rocket engine that provided a constant amount of acceleration. Its speed would increase, so each day we would

The right picture is worth a thousand words

travel further than on the day before. As a result, each successive point on the graph would move up by more than the last one, giving not a straight line but rather a curve bending upwards.

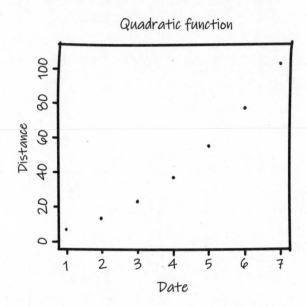

Quadratic function

Technically, this curve is known as a parabola, and this is referred to as *quadratic* behaviour. You can see it in other contexts where acceleration is constant. For example, on throwing a tennis ball across a field we would see the same sort of shape, only bending downwards, as the initial upwards speed of the ball reduced to nothing at its high point, and it then accelerated back to earth.

We illustrate this here, with a variety of curves representing throws of different strengths. It doesn't really matter what the units are on the x-axis and y-axis, but you can think of them as metres if you like. The key is that all these different throws result in the same kinds of parabolas, each one caused by gravity pulling the ball back to earth.

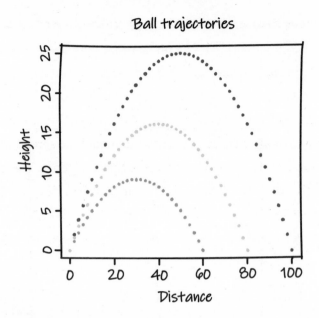

Ball trajectories

Overfitting

More exotic curves are possible too, though these tend not to occur so much in nature. I referred to the parabola as a quadratic because the equation of the curve involves the time squared (multiplied by itself). Although it seems slightly confusing, the terminology 'quadratic' arises because a square is a four-sided shape – think of quadrants and quadrangles to remember it.

In theory, we could think of curves that involved time cubed (time multiplied by itself multiplied by itself), time to the fourth power or even more. Such curves, referred to as *polynomials*, can be deceptive because, given a rich collection of possibilities, a computer can often find one such curve that apparently explains the data for a limited period of time. However, this can happen purely by coincidence, and such curves often fail to predict future values with any degree of accuracy.

The right picture is worth a thousand words

Notoriously, one such curve was produced early in the US COVID outbreak by the former Council of Economic Advisers Chair. Referred to in the *Washington Post* as the 'cubic fit', this model projected US coronavirus deaths 'essentially going to zero by May 15th 2020'. Sadly, it turned out to be extremely far from the truth.

Even more exotic curves were used to model the spread of coronavirus. Isaac Ben-Israel, chair of Israel's Space Agency, gained media coverage for using a polynomial involving time to the sixth power to demonstrate that the virus would inevitably disappear after 70 days. Again, this model did not survive contact with the real world.

As a rule of thumb, it is not enough that a curve should lie close to the data points for a limited period, but rather it needs to make sense why such a curve might occur. For example, there may be some good reason that time cubed or even time to the sixth power might be part of the process that generated the data, but without such a reason it would be prudent to be extremely sceptical of mechanistic curve fitting of this kind.

In general, while it can be tempting to fit overly complicated polynomials to data, there is a danger of what is referred to as *overfitting*. Rather like a dot-to-dot picture, we could find polynomials that apparently perfectly explain a complicated dataset by connecting every point together. However, given more data, it is extremely unlikely that the new points would lie perfectly on this curve. The shape and form of it would be an artefact arising from the original data, with little predictive value for the future.

The great Hungarian-American polymath John von Neumann is said to have remarked, 'With four parameters I can fit an elephant, and with five I can make him wiggle his trunk.' What he meant by this is that given a rich enough class of models (such as polynomials with many terms), it is possible to create mathematical functions which exhibit almost any kind of behaviour. However, such curves may not say anything useful about the real world. Indeed, based on Occam's razor or a desire for simple models in general, we should be suspicious of elaborate explanations, unless there are particularly good grounds to believe that they should hold.

We can illustrate this by returning to the iced bun example. Remember you are a café owner who has sold 143, 136, 147, 144, 149, 147 and 153 buns in successive weeks, and suppose that you would like to predict your future sales. One simple idea would be to fit a polynomial through this data. With some mathematical trickery, it is possible to find a curve which does this *perfectly* – see below. As in the Israeli COVID example, this is a polynomial of degree 6, that is including 'week to the power 6'. As you can see, the curve goes through each of the data points, apparently providing a perfect explanation.

Sales by week

The right picture is worth a thousand words

However, there is a problem in using this curve to forecast future bun sales. As you can see, the curve rises very steeply to week 7, and continues to steepen after that. In fact, if we extend time for one more week, we need to hugely increase the range of the y-axis to even fit the curve on the page. It predicts the café will make 437 bun sales in week 8, which is extremely surprising given the narrow range of values seen so far.

Sales by week

Of course, such a figure is theoretically possible. Perhaps a circus will visit town and place an enormous order on behalf of their performing elephant. Perhaps, like Coventry's previously anonymous Binley Mega Chippy, the café will become the subject of a bizarre TikTok trend and be packed with teenagers from across the country. However, it feels hugely optimistic to assume that will happen simply based on a tame mathematical curve, and to bake that many buns accordingly.

However, even if we believe this first forecast, the curve gets further out of control after that. It suggests that in Week 9 the café will sell 1,907 buns. By Week 15, the projected value is 336,381! This is clearly not feasible. This is a perfect example of overfitting, of drawing a dot-to-dot curve through the points and making foolish extrapolations based on it. It's a curious fact, but models that explain current data too well can be less trustworthy than ones that provide a looser fit.

For this reason, it is always vital to not trust too much in a mathematical model, however superficially attractive it is in the short term, and to subject any forecasts to some kind of 'smell test' to see if they make sense. We will see in Part 2 that one way to avoid overfitting, and to produce forecasts that are both simpler and more plausible, is to embrace the idea of random variation. However, for now we will continue to study more kinds of mathematical structure.

Summary

In this chapter, we have started to think about data, and seen the value of representing it on graphs. This can help us think about how the data might be generated, and to consider the possibility of linear or quadratic growth, for example. However, overfitting may cause us to make highly confident but extremely inaccurate extrapolations of data, and we should always try to think about the underlying processes behind the numbers.

Suggestions

If you would like to try out some of the ideas from this chapter, then it might be fun to look for graphs in the news and in other

The right picture is worth a thousand words

places. Have a think about what story they are trying to tell, see if you can spot any sleights of hand in the way that the numbers are being presented, and think whether it seems reasonable to use them to make informal predictions about where the numbers might go next.

Chapter 2
Getting it in the right ballpark

Comprehension through approximation

At the time of writing (June 2022), the United States national debt is quoted at $30,536,360,095,124. That's a crazy figure even to look at. We are used to there being big numbers in science – even our closest star, Proxima Centauri, is estimated to be 40,208,000,000,000km away for example, and that's just the beginning of the vastness of space – but it's still a jolt to see such huge figures in earthbound human activities.

Each day, you might hear enormous numbers involving thousands, millions, billions or even trillions of pounds; the latest football transfer news, details of big-ticket government spending items, or the size of the national debt. UK members of parliament are paid £84,144 per year. There are around 1.26 million unemployed people in the UK. Elon Musk offered to buy Twitter for $44 billion. In January 2022, Apple became the first company valued at over $3 trillion.

Numbercrunch

It's hard not to just let your mind gloss over this stream of numbers, but of course it's important for us as responsible citizens to try to comprehend what these values actually mean. In this chapter we will see some tricks to help us cope with this rush of numbers and to make sense of it all. This is a valuable skill with applications in many scenarios, given this constant barrage of reported statistics.

First, as I have already mentioned, *any* number should be thought of as somewhat inaccurate, and we should abandon false precision when reporting or thinking about them. For example, Wikipedia tells me that 'As of November 8, 2018, the United States is estimated to have a population of 328,953,020'. The obvious question is, how do they know? Given the constant turnover of births and deaths, immigration and emigration, how could you possibly quote a figure to that degree of accuracy?

Indeed, even if you did somehow know the population to the last digit of precision, what difference would it make to your understanding of anything? Looking at that figure, I would be tempted to round it to 330,000,000 (that is 330 million), or even to 300,000,000. In fact, as a mathematician, I might round it to 333,333,333, and think of it as a third of a billion, but of course I appreciate you may not feel the same way! When estimating things to do with the United States, such as the size of their army or the rate of deaths caused by gunfire, any of those figures would be fine.

Indeed, unreasonably precise predictions or quantifications of data, quoted with a margin of error that is narrow or non-existent, may even be a warning sign. For example, when the polling data website FiveThirtyEight.com gave Joe Biden a 69% probability of winning Florida in the 2020 US presidential election, what did that mean? How does that differ from a 70% probability? In fact, you would have to run the election thousands of times to reliably detect a difference between these scenarios. This degree of spuriously quoted precision was at least an improvement from 2012, when FiveThirtyEight gave Barack Obama a 90.9% probability of winning the election: the difference between this and a 91% probability is unmeasurable in practice.

However, there is a greater problem than this: simply that many numbers involved in news reports and official statistics are just unimaginably huge. Most people might feel comfortable with a thousand or a million, and what they represent, but billions and even trillions are another matter. For clarity, when we talk about a billion, we mean a thousand million (the so-called British billion was abandoned as official government terminology in 1974, and I am unconvinced that it survives anywhere outside old books). A trillion is a thousand billions, or a million millions.

Still, for anyone hearing numbers of this kind, a natural reaction is for their eyes to glaze over, and for them to pay insufficient attention. Perhaps for this reason, and since journalists can be affected by the same phenomenon, it is not unknown for mistakes to appear in the media where figures that should be in billions of pounds are misquoted as the same number of millions, or vice versa. Indeed, in one particularly egregious March 2020 example, MSNBC anchor Brian Williams unquestioningly read out a tweet claiming that, since Michael Bloomberg's unsuccessful 2020 presidential run had cost $500 million and the population of the US was 327 million, Bloomberg could have given each person over a million dollars instead.

Such errors can fatally undermine the credibility of a commentator and lead to humiliating viral clips. For this reason, it may be useful to have some rough comparators in your mind to help check your sums. For example, the most expensive football transfer of all time was Neymar's move to Paris St Germain, for just under £200 million – equivalently, five Neymars make a billion pounds. In contrast, the two newest UK aircraft carriers together cost around £8 billion. The annual NHS budget is around £130 billion. The UK's gross domestic product is around £2 trillion, or about ten thousand Neymars. (All these figures are only approximate and will undoubtedly become out of date: but as described above, we are not looking for spurious precision, but instead some understanding of what quoted numbers mean.)

Hence, if you hear someone claim that the interest on the UK national debt is only £60 million per year, you should be sceptical: is it likely that we owe so little that the interest amounts to the price of a Premier League striker? Similarly, if you hear that a new building would cost £30 billion, you might reasonably ask whether it would be several times more expensive than the jewel in the crown of the British navy. In both cases, it is likely that millions and billions have been swapped. By thinking in terms of these comparators, and becoming familiar with what they mean, you can quickly spot issues like this. It is remarkable how many people will happily accept numbers at face value without thinking whether they pass a basic 'smell test'.

Of course, transposing millions and billions is an extreme error that can be spotted relatively easily. A subtler type of error is quoting figures that are wrong by what mathematicians call an order of magnitude. By this, they mean a factor of 10 – in this sense, two orders of magnitude are 100 (10 times 10), three orders of magnitude are 1,000 (10 times 10 times 10) and so on. Such mistakes can easily arise even with the use of a calculator, simply by typing in the wrong number of zeroes, and unfortunately can be easily missed by an author or editor. This is another reason it is important to develop intuition regarding large numbers – if a value quoted seems too high or too low, there may be an order of magnitude error somewhere, and it's good to double-check.

Understanding budgets

One useful way to understand the large sums of money that arise in government announcements is to think of them on a 'per head' basis. Obviously, the government's spending is not exactly like a household budget, and different people and organisations pay different amounts of taxes, so contributions will not be shared equally. However, we can still do the kind of rounding and rough calculation that I have described above.

As I have mentioned, we can think about the population of the United States as about 300 million or a third of a billion, whichever makes the maths easier. For example, Wikipedia quotes the price of a Gerald Ford class aircraft carrier at $12.8 billion. Again, we can round this, say to $15 billion, because we don't believe that government programmes always cost exactly what was budgeted. So then, $15 billion divided by a third of a billion people gives $45 each (dividing by a third is the same as multiplying by three). Not cheap, but perhaps not ruinous either.

Returning to the number from the start of the chapter, the

Numbercrunch

United States national debt is quoted at $30,536,360,095,124. We'd like to try to make sense of this. First, you need to count digits: 6 zeroes make a million, 9 zeroes make a billion, so let us think of it as $30,536 billion. (Some people will think of it as $30 trillion, but it is best to keep the units the same.) Again, we can ignore the spurious precision – the last few digits are already out of date – and think of this as $30,000 billion, which divided by a third of a billion people gives $90,000 each. That's more than most people will earn in a year.

Of course, if we'd not rounded any of these numbers, and believed they were exactly accurate, we could have done the same calculation – probably not in our heads, but we could use a calculator. If I do so now, then I find the cost of an aircraft carrier is actually $38.91 each, and the national debt is $92,282 per head. Knowing these answers to greater precision doesn't really help my understanding.

None of this is meant to form a judgement on the worth or cost-effectiveness of any single government programme, but it is invaluable being able to perform these approximate calculations to break down the headline figures into something more relatable. Certainly, it can allow you to be sceptical if someone suggests that the US could pay off its debt by cancelling an aircraft carrier. Indeed, it is striking how often Parkinson's Law of Triviality holds, in the sense that discussion is often disproportionately focused on small sums of money, rather than seeing the bigger picture.

It's good to practise these kinds of calculations, and to get into the habit of thinking about numbers in the news in this way. As a result, you might find yourself more sceptical about claims you read on social media. I suggest that you practise now by trying a few of these kinds of sums. If you read on Facebook that we should raise unemployment benefits by halving MPs' salaries, it's a nice emotive point, but how much would that raise for each unemployed person? Similarly, how much does the average person pay in taxes towards the Royal Family each year?

Indeed, on a similar principle of rounding the numbers to make the sums easier, it is useful to bear other figures in mind. The population of Scotland is about 5 million, there are about 9 million people over the age of 70 in the UK, there are about 30 million UK households and 2 million students, and so on. None of these figures are exact or up to date, but they allow us to calibrate the effect of proposed measures relative to their costings.

One further caveat is that sometimes announcements of this kind involve multi-year programmes: I don't imagine the United States buys a new aircraft carrier every year, for example. In that case, as a first guess we might want to simply divide the headline figure per person per year. However, as we will see in Chapter 3, money and compound interest are an example of where so-called exponential growth occurs, so it may be worth mentally inflating some of these figures somewhat to account for overspend over time. Equally we should remember that a given sum of money will not be worth the same in 2040 because of inflation which has an effect like exponential decay (again see Chapter 3 for more about this) in terms of 'what a billion will buy us'. These issues partly explain the difficulty of governments making long-term budgetary forecasts.

Fermi estimation

This kind of reasoning is an example of an interesting way of thinking, which it can be invaluable to practise. It is known as *Fermi estimation*, named after the great physicist Enrico Fermi. Fermi's most famous feat of this kind was estimating the explosive yield of the first atomic bomb test in July 1945, by dropping pieces of paper and watching how far they were blown by the blast wave. Knowing roughly how much they were blown and from approximately what height they'd been dropped, he could estimate the pressure exerted by the bomb's shock wave. By estimating roughly how far he stood from the detonation, he could work out how much energy must have spread out from the original blast to cause that much pressure at such a distance. Remarkably, Fermi's crude methods enabled him to come up with a simple estimate that was within a factor of two of the final accepted value. Similar methods were applied after the Beirut port explosion of August 2020, to estimate the size of the blast from video footage of a bride's dress being blown by the shock wave.

It is worth understanding how the method works. A classic example is the job interview question, 'How many piano tuners are there in Bristol?' One could just guess, but a better way is to use Fermi's method and break it into stages. How many people are there in Bristol? What percentage of people have a piano? How often does a piano need tuning? How long does it take? How many hours per day and how many days a year do people work? Each of these quantities seem more reasonable to estimate than simply guessing a final figure in one go, and they can be combined to get a sensible overall answer to the problem.

Getting it in the right ballpark

For example, I might guess the population of Bristol is 500,000. Maybe 2% of people have a piano, which means there's about 10,000 pianos to tune. Perhaps a piano needs tuning once a year, and it takes an hour each time, so that's about 10,000 hours of piano tuning needed yearly. Maybe one person works eight hours a day, 200 days a year, so one tuner could do about 1,600 hours of tuning. So maybe we need about six piano tuners? If I look on Google, it seems there are about nine or ten in Bristol. I wasn't right, but it wasn't a terrible answer either.

The magic of Fermi estimation comes through the way that these different estimates are combined to answer the whole problem. Clearly, we do not expect each individual estimate to be perfect. However, we find the overall answer by multiplying the various estimates together in the right way. We might reasonably expect each sub-estimate is equally likely to be too high or too low, and as a result the errors tend to cancel each other out, and the eventual answer that we derive can be surprisingly accurate.

We can understand this better in terms of the Law of Large Numbers which will be discussed later in Chapter 5. Informally speaking, this result tells us that randomness and error 'tend to average themselves out', if the errors are made independently and enough random terms are combined together.

This suggests that, within reason, the more stages into which we can divide a Fermi problem the better. It is reasonable to believe that the estimates at each stage should be independent since misunderstandings of one data source are unlikely to filter over to another. For example, the problem of estimating how many people live in Bristol is totally unrelated to the question of estimating how long it takes to tune a piano. For this reason, Fermi estimation often performs extremely well as a tool to gain a quick and approximate first answer to complex problems.

Another famous example of Fermi estimation is given by the so-called Drake equation, which seeks to estimate the number of planets in the Milky Way with a level of civilisation sufficient to generate signals detectable from Earth. Again, this proceeds in a similar multiplicative fashion: considering at what rate stars are formed, what proportion of those have planets and so on.

However, the resulting answer indicates one major limitation of Fermi estimation. Depending on the values of these sub-estimates, it is possible to arrive at a value anywhere between a millionth of a millionth or tens of millions for the number of such planets. The reason for this wide variation is the considerable uncertainty regarding the value of some of the terms that Drake uses.

For example, his equation includes the average lifetime of a civilisation that is sophisticated enough to radiate signals into space. Drake himself thought this could be anywhere between 1,000 and 100,000,000 years – already a factor of 100,000 difference. Another term is the proportion of planets where life evolves which go on to develop intelligent life. Estimates for this vary between a billionth and 1, giving another factor of 1,000,000,000 difference. These uncertainties will be multiplied together, already combining to give a factor of 100,000,000,000,000 difference in the eventual answer.

The problem is that these terms are essentially unknowable. We have little evidence in either direction to know what life on other planets could be like, compared with our lived experience which lets us guesstimate plausible figures for what proportion of Bristolians own a piano. We might conclude that the Fermi method works best when we have reasonable grounds to estimate the answer to each problem, which is far from the case for the Drake equation.

Approximation and Infection Fatality Rate

The skills of numerical reasoning, estimating a plausible figure and checking whether quoted numbers pass a 'smell test', were extremely valuable during the coronavirus pandemic. Two numbers that are worth discussing in this context are the *Infection Fatality Rate* (IFR) and the *Herd Immunity Threshold* (HIT). Thinking about them together allows us to check some especially outlandish claims. Further, by illustrating the caveats and issues that arise even with these simple numbers, we can see how careful you need to be with reported numbers in any context.

The Infection Fatality Rate of a disease is the percentage of infected people that will die. However, it is important to understand that a disease like COVID-19 does not have a single IFR – not all people were affected equally, with particularly notable differences of risk by age group. For concreteness, I will give some figures for the UK, estimated by the Medical Research Council Biostatistics Unit (MRC-BSU) in the summer of 2020.

It is extremely striking how much more severe the illness was for the old: the MRC-BSU estimated a fatality rate of 11% among the over 75s, falling to 2% among 65–74-year-olds, and below 0.04% for all age ranges under the age of 44 (with even lower values for the young). This led to an overall estimated UK average IFR of 0.7%. This sharp disparity in outcomes, known since the early days of the pandemic, explained the need to protect the old from infection and motivated the plan to vaccinate those age groups first, even bearing in mind that the death of a younger person has a greater impact in terms of years of life lost.

Even beyond that, it is worth remembering that the IFR depends on healthcare: given hospital treatment some people will survive who would otherwise die. This indicates the danger of hospitals being overwhelmed. Given a shortage of beds, trained staff or other resources, the IFR would undoubtedly rise, as tragically seen in India in spring 2021 for example.

One popular alternative to the IFR is the *Case Fatality Rate* (CFR), which is the proportion of people with a positive test who went on to die of the disease. This has the advantage of being easier to calculate, since both numbers are readily available. However, since case numbers depend on test availability, so will CFR, meaning that figures early and later in the pandemic were not comparable. Certainly, since testing will never capture all the infections, the CFR is not the same as the IFR, and you should not use them interchangeably. Further, there is a question of lag, particularly in an exponentially rising or falling epidemic – we need to compare deaths today with the cases that led to them, which were those roughly 21–28 days previously.

Using age-based IFRs, it was possible to forecast deaths much more accurately than by crude calculations based on case count alone. For example, in France it was reported on 29 October 2020 that there had been 375 positive tests per 100,000 people over 75 in a week. If we call this 50 positive tests per day, a Fermi calculation would suggest: French population is 70 million, 10% are over 75, so 7 million over 75s (70 times more than the sample), so 50 x 70 gives

3,500 cases in that age group per day, which at IFR 10% means 350 fatalities from those testing positive, so if there were twice as many infected people as positive tests that would suggest 700 fatalities per day from that age group alone. Like other Fermi estimates, this figure was not completely reliable, in that three weeks later France reached 626 deaths per day in total, but it was a useful starting point for comparison.

It was striking that during summer 2020, case counts rose considerably in Florida, France and Spain without an immediate rise in fatalities, even taking lag between infections and deaths into account. This led to a narrative of a so-called 'casedemic', and a suggestion that the virus had become less potent. However, as subsequent rises in deaths showed, this was an illusion, caused both by the greater availability of testing (and hence reduction in the CFR) and by the fact that early infections were concentrated in the young before spreading into older age groups.

Approximation and the Herd Immunity Threshold

Another number which attracted a lot of discussion is the Herd Immunity Threshold (HIT). In a conventional model, at the start of an epidemic each infected person will infect a number R0 of people. As time goes on, even without social distancing, this number will eventually start to drop, as previously infected members of the population become immune to the virus. Some of the people who would otherwise have been infected no longer can be, and so the rate of infection gradually drops.

We can even work out how many people we might need to have been infected: if at least a proportion (R0-1)/R0 of the population has been infected, then more than (R0-1) of the contacts that would previously have led to infection no longer do. This means that fewer than 1 person is infected, and the infection will die out (in fact, this will happen at an exponential speed – see Chapter 3 for more about what this means). This ratio (R0-1)/R0 is referred to as the Herd Immunity Threshold. Standard estimates suggested that R0 of the original COVID strain was about 3, so the HIT would be 2/3, or 66% of the population. So, once more than 66% of the population had been infected, the number of cases would inevitably start to fall.

A variety of arguments were put forward to suggest that this figure may be an overestimate. For example, perhaps a proportion of the population had some degree of pre-existing immunity. Other mathematical models, described in more detail in Chapter 10, suggested that people who have more contacts (through their job or living arrangements) were more likely to contract the disease first, and their removal from susceptibility to the disease will have a larger effect than a randomly chosen person.

There may be a degree of truth in both these arguments. However, towards the end of the first wave of the pandemic in Europe, theories emerged that the HIT could be as low as 20%, and that having achieved this level, the disease was naturally dying out everywhere. Subsequent developments show this to have been far too optimistic, and that the decay in infections is likely to have been due to social distancing and the effect of Europeans spending more time outside over the summer.

In fact, simple arguments can rule out some more fanciful figures for IFR and HIT immediately. Since over 196,000 people in the UK out of a population of roughly 70 million have now had COVID recorded on their death certificate, we can immediately work out that the overall IFR must have been at least 0.28%, even if everybody had been infected. In fact, it is likely that vaccinations had lowered the United Kingdom's IFR by early 2022 – with estimates closer to 0.05% for example – but clearly the original strain (and the deadlier alpha and delta variants) had a higher fatality rate than many people claimed, particularly among the unvaccinated.

Further, we can put the IFR and HIT together to show that, despite what you may read from certain commentators, it is literally impossible to be simultaneously too optimistic about both, because they pull in opposite directions. For example, if someone claims that the virus was not serious, because the IFR is 0.2% and HIT is 20%, then at most 14 million people in the UK could have been infected, and at most 28,000 of those would have died. The claim is impossible, based on the data that we have already, and can be rebutted instantly by anyone with a calculator. We can even debunk these figures using the approximate method that I have described, because the number of deaths entailed is so far below the observed data that the calculation has a comfortable margin of error.

Counting emails

If we all had a better feel for numbers, false claims regarding the pandemic and other issues would not circulate as readily, and the tricks I have described can help you achieve this instinct. In general, it is important to understand reported data and survey statistics, which can often be reported in a selective way to bolster a particular conclusion.

Numbercrunch

I will give one more example to show how thinking about numbers in this way can help make sense of everyday life. Like most of us, I feel besieged by email. I have several accounts, both personal and professional, including shared inboxes to monitor as part of my job. But am I unusual in this? What is a normal number of emails to send or receive each day?

As often with this kind of question, the answer may well depend on what you mean, but trying to pin down the details can often help you understand more than just hearing a definitive answer would do. If I Google 'daily worldwide email traffic', I am pointed at a page from statista.com, which suggests that 306.4 billion emails were sent each day in 2020. I don't have any reason to doubt this, so we can take this as an accurate figure.

Using this number, we can start to think what a normal inbox might look like. As usual, we will round it to 300 billion, and think about it on a per capita basis. The population of the world is around 7.7 billion, so if I round that to 7.5 billion then I get a nice simple answer: the number of emails divided by the number of people in the world is 40. So, naively perhaps we are all sending and receiving 40 messages a day on average. Quite a lot, but not unmanageable perhaps?

However, we need to think carefully about what we mean by an average and consider whether that is a representative measure. Again, Googling 'how many people in the world use email' suggests that there are only 4 billion users worldwide, or just over half the world's population. So perhaps there is an argument that we should divide the 300 billion emails among the active users, to produce a figure of 75 messages each, which starts to feel a bit more daunting?

Getting it in the right ballpark

There's not a right or a wrong answer as to whether we do this, but it's worth at least appreciating that it makes a difference.

Further, thinking about it more, we can see that there may not even be a single figure for email use, as this headline figure might suggest. That is, email is a fundamentally asymmetric means of communication. A message goes in one direction only and does not guarantee a reply. For example, if a supermarket contacted all their UK customers with an important announcement, that might easily be a communication to 5 million people in one go. Maybe some of them would reply, but the majority wouldn't. It starts to become clear that typical email levels could be quite different according to whether we consider incoming or outgoing messages.

This also indicates another way in which the headline count of 300 billion is unclear. If I send an email which is copied to six people, does that count as six messages or as one? In terms of effort for me to type it should only count as one, but in terms of total cognitive overload for the receivers it should count as six. I will assume that it counts as six, but it's not completely clear whether that's how I should understand the 300 billion.

Since I have decided that an email sent to six people should count as six separate messages, we can see that the total number of incoming and outgoing messages should be the same, and therefore the sample average of the two types of messages will be the same. However, this average does not capture the whole story.

For messages sent, there would be a substantial number of extremely heavy traffic accounts – shops, newsletters, spam accounts run by bots – essentially accounts that operate in broadcast mode, sending messages to large numbers of people. As a result, it's quite plausible that I personally send fewer emails than the average, which is inflated by these heavy hitters.

On the other hand, for messages received, the range of values will be much smaller. There are likely to be fewer accounts with extremely high incoming traffic – even the customer service address for a supermarket would not hear from all its customers in one day. I'm on a number of mailing lists, and so I wouldn't be surprised that my incoming traffic is close to or even above the average.

All this seems intuitively reasonable. I generally send fewer emails than I receive, even accounting for the messages that I cc to multiple people. However, the incoming traffic takes less mental effort than writing outgoing messages, since much of my inbox consists of mass communications and newsletters which can be easily ignored, filtered or deleted, so my feeling of overload may not be down to raw numbers alone.

It is worth noticing that none of this discussion really answers the original question of whether my levels of email are unusual. Ultimately, the general wider context and overall global averages may not be relevant. I am not a bot generating spam messages, nor do I live in the Kalahari Desert with no access to the Internet. The only reasonable context is that I am an office worker in a Western country, and so should compare myself with that cohort of people.

The best way to do that might be not to look at overall averages at all, but rather to perform a targeted survey, rather like an opinion poll. Even that would come with caveats, not just in terms of the incoming and outgoing issue I have mentioned above, but also in terms of sampling (which office workers do we choose and how?) and reporting (do you simply ask me about my email traffic, which is prone to inadvertent exaggeration, or do you monitor it directly?). I will return to these issues in Chapter 11.

Overall, with a question like 'is my email use normal?', the message is that sometimes the right answer isn't a single number, but rather 'it depends on what you mean by that'.

Summary

In this chapter, we have seen several ways to make sense of numbers that you might see in the news or elsewhere. It's fine to make approximations, and tricks like dividing scary-sounding budget items up on a per-head basis and using Fermi estimation can help us navigate our way through the jungle of numbers. I've illustrated these ideas with some examples, including the question of IFR and HIT from the coronavirus pandemic, and shown how hard it can be to give a single definitive answer to a question like 'how many emails do people receive?'.

Suggestions

I encourage you to practise the tricks that I have described, until they become second nature to you. When you see big numbers in the news or elsewhere, try thinking about what they really mean, and use some of the tricks I've described to kick the tyres a little bit. Who knows, you might even spot an error in a journalist's or a politician's calculations – if you do, then I'd love to hear about it! You might also like to try counting your incoming and outgoing emails every day for a month, to see how your numbers match up.

Chapter 3
Captain's log

Football transfer fees and exponential growth

You may have a sense that football transfer fees are completely out of control. It is not unusual to hear of players changing hands for £50 million, £60 million or more – as I've already mentioned the record fee is the £198 million paid for Neymar in 2017. In comparison, the first ever transfer record was the princely sum of £100 paid in 1893 by my own team, Aston Villa, to buy the Scottish forward Willie Groves. It seems like a crazy jump, to get from Groves to Neymar in just 124 years. What can we expect next? Is a billion-pound footballer a reasonable possibility?

As I've already described, a sensible way to gain an initial understanding of data is to plot it on a graph. It's fairly easy to download the successive transfer record from Wikipedia, and to visualise how it has changed over time.

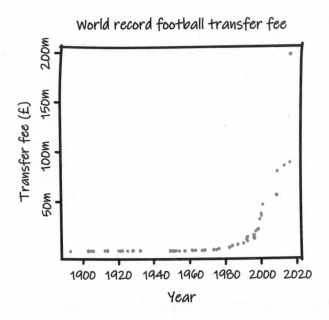

World record football transfer fee

The picture seems clear and appears to confirm our impression that transfer fees have grown out of control relatively recently. In fact, it seems like the record hardly changed up until 1980, and that all the growth has come since then – indeed even up to the year 2000 the record hardly moved at all.

However, we will see in this chapter that this perception is misleading and is really an artefact of the way in which the data has been plotted. In fact, we are seeing an example of *exponential* growth, and we can understand the progression of the transfer record much better by using a different kind of graph, with what we call a logarithmic scale.

While the polynomials that we saw in Chapter 1 are interesting and useful, this different exponential curve can predict the future of certain phenomena extremely well. We will see that exponentials arise in biological problems such as the growth of bacteria or human populations, in problems in finance and economics to do with the growth of money over time, and in the performance of new generations of computers. By understanding exponentials, we will understand all these processes much better, and by plotting them in the right way we will be able to extrapolate their growth more easily.

Exponential functions can be described in a deceptively simple fashion. In contrast to our Voyager space probe from Chapter 1, where each day we *add* a fixed amount to the distance travelled, with exponentials each day we *multiply* the previous value by a fixed amount. Moving from addition to multiplication sounds like a tiny change, but this apparently innocuous difference transforms the tame linear behaviour we saw before into something much more pathological and potentially harmful.

The classic example of an exponential process is that of a bacterium which divides in two parts on a regular basis. For example, if each bacterium divides in two every hour, then we can compute the size of the population very easily. If at midnight there is one bacterium, at 1am there will be two, at 2am there will be four and by 3am there will be eight. So far this doesn't seem too worrying and indeed, like the transfer fee graph, the initial stages of exponential growth can seem flat or be mistaken for linear or quadratic growth – but only for a while.

Things would go wrong extremely fast. By 8am, after five further doublings there would be 256 bacteria. By midday there would be 4,096. By the next midnight there would be 16,777,216 bacteria. Having started in a way that can be hard to distinguish by eye from linear growth, the exponential growth has rapidly outstripped linear behaviour. This is a universal feature of exponential processes: it can be proved mathematically that any amount of exponential growth eventually overtakes any rate of linear growth. Indeed, exponential growth will overtake any of the polynomial processes (quadratic, cubic or sixth power) described in Chapter 1.

This is another reason such mechanistic reliance on polynomial processes in the modelling can be deceptive and dangerous, by lulling observers into a false sense of security. A particular danger is that an overwhelming proportion of the overall growth takes place in the later stages of an exponential process. This means that, in a pandemic, hospitals can rapidly go from under control to overwhelmed apparently out of nowhere.

A nuclear chain reaction is another instance of exponential growth which may help us understand this. If an atom of uranium-235 is bombarded by a neutron, it will break apart into pieces, releasing more neutrons and a certain amount of energy. For example, if it releases two neutrons then those can in turn collide with more uranium atoms, producing twice as much energy. At the next generation, there would be four neutrons released, and four times as much energy and so on. In contrast to the bacteria, these generations may only last for a few millionths of a second, meaning that the amount of energy released can become enormous almost instantaneously. This chain reaction is precisely the principle behind the atomic fission bomb dropped on Hiroshima in 1945. By providing a large enough sample of uranium, and by designing the shape of the bomb casing in such a way that few neutrons can escape, the amount of energy released can grow exponentially for long enough to flatten a city.

Captain's log

Of course, unchecked endless exponential growth is more likely to be encountered in the world of mathematics than in the wild. It would be foolish to assume that growth would continue in this way indefinitely because in practice there will always be limits on the size of the process. For example, the bacterial population may be constrained by the capacity of its container, or by the availability of chemicals to build the bacteria themselves. Similarly, the size of the sample of uranium will constrain the length of time over which a nuclear chain reaction can occur, limiting the energy released by a particular fission bomb. However, the COVID pandemic demonstrated repeatedly around the world that exponential growth can keep going long enough to cause serious problems for healthcare services.

Exponential growth remains widely misunderstood. One way in which this manifests itself is through an incorrect assumption that the only kind of exponential growth possible is of the doubling kind described above. However, there is a whole family of exponentials. The best way to understand this may be through a more familiar object: your bank account.

We are used to the idea that money held in the bank can earn interest, and that money borrowed from the bank will be charged interest. This happens in a way that involves multiplication which, due to the phenomenon of compound interest, creates exponential growth. For example, if we borrow £1,000 at 10% annual interest, after a year an amount of £100 would be added to the debt, meaning that we now owe £1,100. The next year, the interest due would be 10% of that amount, giving £110, meaning that we now owe £1,210, and so the following year's interest would be 10% of *that* amount, giving £121. Each year's interest includes interest owing on the previous years' interest, so the increases in debt get larger each time.

This is the same kind of exponential growth that we saw with the bacteria example. In general, exponential behaviour emerges whenever a quantity is multiplied by the same constant factor in each period, and any fixed rate of annual interest leads to exponential growth. Whether the interest rate was 1% or 50%, the debt would follow a curve that started off relatively flat and became steeper with time. Indeed, the debt would eventually exceed any credit limit, whether a million or a trillion pounds. The only difference between these scenarios is how long it would take to cross this threshold, which is determined by the rate of interest.

Similarly, we can think about the nuclear example again. By controlling this kind of chain reaction more carefully, it can be sustained at a steady level, providing the basis for nuclear power plants to operate. That is, by placing control rods into a nuclear reactor, we can absorb a constant fraction of the emitted neutrons, meaning that on average each reaction releases one neutron which escapes absorption. This is referred to as a critical reaction: in theory it can remain at this steady pace, emitting a constant amount of energy.

It is slightly strange to think that this kind of tame constant function also represents exponential behaviour. However, just as a constant function can be obtained by adding zero at each time step, it can also be obtained by multiplying by 1 each time. The issue is that, without care, this kind of critical chain reaction could slip out of control. If the control rods were not placed correctly and too few neutrons were absorbed, the energy released will be multiplied by a number greater than 1, and the kind of destructive exponential growth described above would be a real danger.

One useful idea in this context is that of *doubling time*. That is, since all exponentially growing processes will eventually pass every level, we can ask how long it would take for them to move from value 1 to value 2, say. Since the process is multiplied by the same factor in each period, it would take exactly as long to do this as it would to move from value 100 to value 200, or from value 37 to value 74, say.

Thinking about doubling time is a simple way to characterise the rate of growth of an exponential process. For example, the bacteria example above has a doubling time of 1 hour – between any time and the corresponding point 1 hour later, the size of the population will have doubled. In contrast, a bank debt at 10% interest would take between 7 and 8 years to double in size.

Log scales

Although I have argued that it is valuable to plot data on a graph, you run into a problem when plotting exponentials. Specifically, all exponentials look superficially the same because the numbers involved grow so fast. If we consider the first few hours of the bacteria example, we will produce a plot that looks like the transfer fee example, staying very flat for a long time before jumping up rapidly at the end. This makes the future behaviour hard to extrapolate by eye. Further, other exponential processes growing at other rates have a similar shape, making it hard to tell them apart just by looking at graphs.

However, one simple data plotting trick can make exponentials

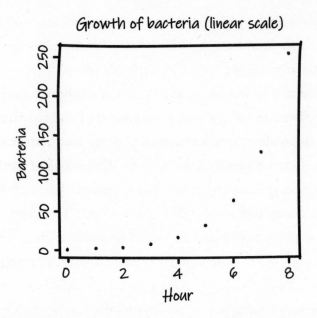

easier to understand. This is referred to as a logarithmic scale, or *log scale* for short. I have become increasingly convinced that many arguments regarding the recent COVID pandemic would have been resolved if more news organisations plotted daily coronavirus data using a log scale and have tried to evangelise for them whenever I can.

So, what is a log scale, and what does it tell us? You may remember that I referred to the vertical axis of a graph as the y-axis. A log scale simply takes the y-axis and squashes it in a particular way. Those people old enough to have used slide rules at school may remember a mathematical object called a logarithm. Essentially, logarithms get rid of exponentials, by turning the multiplication that generates them back into the tamer operation of addition.

For example, think of the fact that 8 x 4 = 32. We can rewrite this in terms of powers as $2^3 \times 2^2 = 2^5$. Just looking at the powers, you will see numbers that crop up in a sum involving addition, that 3 +

2 = 5. This is no coincidence. In fact, we call 3 the logarithm of 8.[5] That is 3 is the number that, if you take 2 to that power, will give you 8. Similarly, 2 is the logarithm of 4 and 5 is the logarithm of 32. The golden rule is that the logarithm (5) of the product (32) is the sum of the logarithms (3 + 2) of the numbers (4 and 8) we multiply.

On a log scale graph, instead of plotting the numbers themselves, in effect we plot their logarithms. If you look at the following graph, you will notice the y-axis has strangely positioned labels. This is an example of a log scale. The size of the gap from 1 to 2 is the same as the gap from 5 to 10, and from 10 to 20, from 50 to 100 and from 100 to 200. This is because each of these steps represents a doubling, which corresponds to adding a fixed amount on to the logarithm each time. We can see what effect this has by replotting the same numbers from the bacteria example on a log scale.

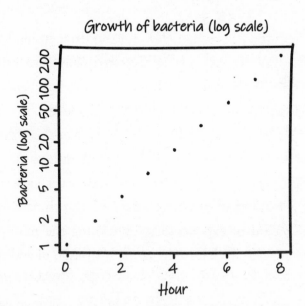

Growth of bacteria (log scale)

5 Technically, it is the logarithm 'to base 2' since the calculation involves powers of 2. If we considered 8 as a power of other numbers, we would get a different answer.

Treating the y-axis like this turns hard-to-visualise exponential growth into something much easier to understand: a straight line. Essentially, the fact that the exponential process is multiplied by the same factor each time is equivalent to the fact that the logarithm of the exponential process has the same amount added to it. (Don't worry if you can't see why this is true!). Notice that this looks exactly like the graph of the Voyager example, so if we can understand that, then we can understand this.

The key idea is this: taking the logarithm of an exponential process gives a linear process. The more steeply the line goes up, the faster the process is growing. Hence, we can compare two exponential processes by plotting them on the same log scale and seeing which has the steeper slope. We can find the doubling time by seeing how long it takes the line to go up from the marks from 5 to 10, or from 10 to 20. Further, turning the rapidly steepening exponential curve into a predictable straight line means that it is much easier to extrapolate future values, with the labels on the y-axis guiding you as to what to expect.

We can now see that, as mentioned at the start of the chapter, the football transfer record has tended to grow exponentially over time. Using a log scale, we obtain the following plot.

We can see several interesting things. Firstly, our impression that there was little growth up until 1980 is entirely wrong. Growth did occur, only not in the sense we are used to. At that time, the record transfer was Paolo Rossi's £1,750,000 move to Juventus in 1976. In an additive sense, there seems a much smaller jump from Groves to Rossi (an increase of £1,749,900) than from Rossi to Neymar (£196,250,000). However, if we

Captain's log

World record football transfer fee (log scale)

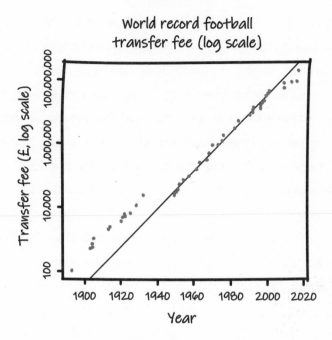

think multiplicatively, as we should, then the jump from Groves to Rossi (by a factor of 17,500) is much larger than from Rossi to Neymar (just over 113). Since exponential growth means that values tend to be multiplied by the same factor each year, this need not surprise us, since more time elapsed between 1893 and 1976 than between 1976 and 2017.

Further, we can draw a straight line on the graph that closely fits the points from 1945 to 2000. Throughout this time, there was approximately exponential growth – the record fee kept increasing by roughly 15% per year. Indeed, we can see that a parallel version of the line would fit the points from 1893 to 1940 as well. It seems that there was exponential growth then too, at a similar rate, which was halted by the Second World War, and resumed later.

Indeed, we can see that our impression of runaway recent

transfer growth is not entirely correct. In fact, the five most recent points lie below the line that projecting exponential growth forward might suggest. After Zidane's move in 2001, there was a relatively long pause until the move of Kaka (in 2009), followed by Ronaldo, Bale and Pogba, which were all for considerably lower amounts than our 15% annual increase might suggest. And while Neymar's move was a steep increase (doubling the previous record), the final point still lies below the line, which would have suggested a £360 million footballer by then. In that sense, it is possible to argue that actually Neymar was undervalued, in the sense that his transfer fee was less than a naive extrapolation of late 20th-century exponential growth might have suggested. Perhaps like our bacteria sample would, this exponential growth has reached some hard upper limit, at least for now.

Incidentally, one often overlooked fact is that there are also exponential processes that get smaller with time, sometimes referred to as exponential decay. In the examples I have described so far, the process grew because we multiplied it by a number that was bigger than 1 in each time step. However, there is no reason that this should be the case. We could instead multiply the process by a number smaller than 1 each time.

One scenario where this occurs is when studying radioactive decay. Given a sample of radioactive material, in each set period a fixed proportion of the material undergoes a nuclear change. Hence at each time, the amount of remaining material of the original type is multiplied by a number smaller than 1, and its total mass undergoes exponential decay. Of course, we can't talk about doubling time any more since the process is shrinking. Instead, we can refer to halving time. In the radioactive decay example, this is known as the half-life of the material.

In the same way, exponential decay would occur in a nuclear reactor where the control rods absorbed too many neutrons. We would multiply the energy released by a number less than 1 each time, and the reaction would rapidly fizzle out.

We don't need any new data visualisation tricks to plot a decaying exponential process – we simply use a log scale as before. The key thing is that the logarithm of such a process has a fixed amount subtracted from it in each time step, just like thinking about Voyager from the alien's viewpoint. As a result, plotting a decaying exponential process on a log scale gives a straight line sloping downwards, exactly as in that case.

Exponential growth and pandemics

Using all this, we can discuss the spread of infectious diseases in mathematical language and understand the right sort of plots to use. The key concept here is the *R number*: how many people each infected individual themselves infects. If R is constant, the size of the infected population will be multiplied by that factor between each generation of the virus. Hence the number of infected people behaves exponentially, at a rate determined by R. If the R number is greater than 1 then this infected population will grow and if the R number is less than 1 then it will shrink.

In either case, as described, it is natural to plot the size of the infected population on a log scale. Or rather, it would be if we knew exactly how many people were infected. However, it is likely that for any disease we will never have access to this information and will need to estimate it via imperfect proxies.

For example, when daily coronavirus cases were reported in the media during the recent COVID pandemic, it referred to the number of positive tests, not the number of infections. On a short-term basis though, we can assume the testing system remained equally efficient from day to day, so a roughly constant proportion of all the infected people tested positive each day. If so, we would see exponential growth in cases, and could estimate the rate of increase of infections through that.

Other proxies for the rate of infection suffered from similar issues. While both daily reported hospitalisations and deaths may not have been quite so susceptible to lack of testing availability, we would still need to assume that a fixed proportion of infected people are hospitalised or die to estimate infection numbers from these figures.

Another issue with hospitalisation and death data is *lag*: it may take 10–14 days for infected people to be hospitalised, and 21–28 days for them to die. This means that these other metrics could only give information about the past rate of spread, making it harder to assess the effectiveness of lockdowns or other interventions.

However, despite these caveats, broadly speaking we can use these three measures to estimate the rate of growth in infections. Since we believe that the infected population should be growing or declining at an exponential rate, this means that plotting these quantities on log scale graphs should produce straight lines, each with the same slope.

Furthermore, although the true number of infections remains

unknown, we can use such graphs to estimate the R number itself. The simplest observation is that if R is greater than 1 then we would see a straight line sloping upwards, and R being less than 1 would give a straight line sloping down. It is particularly easy to estimate the doubling or halving time, but to deduce the R number we need to make assumptions about how long it takes for an infection to be passed on.

What we would hope to see is that the R number would reduce over time, for example as social distancing and hygiene guidelines took effect, or as more people became immune. On the log scale graphs, this would manifest itself through rising straight lines starting to flatten, before peaking and then declining. In other words, we might hope to see something like the tennis ball trajectories of Chapter 1.

Using a log scale, we can rapidly assimilate the progression of the COVID pandemic in the UK. Here is a plot of daily deaths between March 2020 and June 2022, which divides into six phases. First, roughly exponential growth, corresponding to R being close to 3, created a steeply rising straight line, flattening a little due to social distancing. This was followed by a second phase of much slower exponential decay (straight line downwards) until early September 2020, as lockdown pushed the R number below 1, and the number of infections shrank. The third phase from September 2020 to January 2021 is more complicated: overall there was a less steep but still consistent phase of exponential growth, as R rose above 1 again. However, it is possible to discern the effect of the second English national lockdown in flattening the death curve for a while, before the arrival of the alpha variant in Kent led to a steepening again. The fourth phase from January 2021 onwards consisted of sustained exponential decay, caused by the third national lockdown and vaccinations. The fifth phase, from early June 2021, again saw exponential growth caused by the delta variant and the relaxation of national restrictions. Finally, since around September 2021, the UK's death figures have been in flux – with relatively short periods of exponential growth and decay caused by the arrival of further variants and booster campaigns, but little overall direction in the trends.

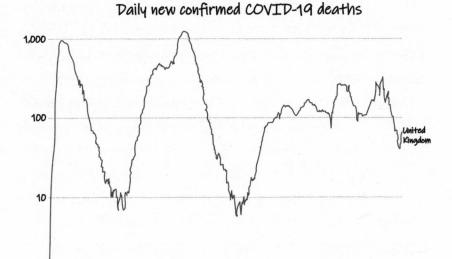

Daily new confirmed COVID-19 deaths

One important remark is that exponential growth of infections will not go on forever, just as in the bacteria example above. Indeed, the famous SIR epidemic model which we will see in the next chapter shows that by itself a growing proportion of immunity begins to flatten the straight lines on log scale plots. For example, in a scenario where 25% of the population had immunity through infection, then a quarter of the contacts that would have previously led to a new infection would no longer do so, forcing down the R number.

In fact, once a large enough proportion of the population had been infected, the R number would be reduced below 1 by this effect, meaning that the epidemic would naturally reduce in size. This is the effect that is being referred to when people discuss the notorious Herd Immunity Threshold. However, conventional epidemic models suggest that this needs a substantial proportion of the population to be infected.

Exponential growth and the stock market

Other everyday processes also exhibit exponential growth, particularly in the world of finance which is built around multiplicative change. Over the long term, compounded annual interest or

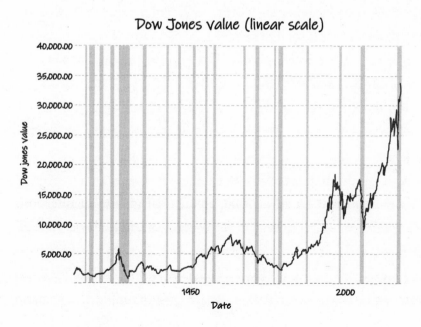

Dow Jones value (linear scale)

inflation rates of a few per cent can make a major difference to the value of investments. As a result, it is natural to think about the long-term growth of the values of companies and assets using a log scale. For example, we can consider the inflation-adjusted performance of the Dow Jones industrial index over about a century. Plotted on a linear scale it looks like the graph on the previous page.

We can see that it is much 'wigglier' than the toy bacterial models or the simple epidemic curves we have seen so far. There are notable fluctuations going on, both on a month-to-month scale (small perturbations of the line), but also more major changes of direction on an annual or five-year basis. I will talk more about the reasons for this in Chapter 10.

Roughly speaking the Dow Jones index seems to grow exponentially, somewhat like the football transfer record. It is very flat for an extended period, before suddenly taking off sometime in the mid-1990s, and steepening further in recent years. Since we know that financial assets tend to behave exponentially, it is a natural idea to plot this index on a log scale. If we do this, again the picture is transformed into something more useful.

Representing the data this way, a new picture emerges. In fact, there is roughly a steady linear rate of growth across the whole time period (albeit with some setbacks). On this log scale, the huge growth in absolute numbers from the 1990s onwards is seen to be a continuation of an ongoing trend, and we can compare the effect of events such as the 11 September attacks and the 2007/08 financial crisis.

Further, some events that seemed insignificant on the linear scale, such as a sustained falling back of the inflation-adjusted index

Dow Jones value (log scale)

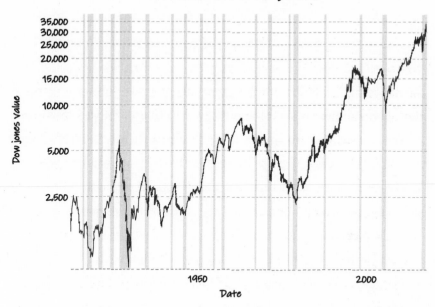

in the 1970s caused by large price rises, played a much greater role in the long-term evolution of the Dow Jones index than the simple linear plot suggests.

In the short term, effects of inflation and growth may not matter much, and it would be fine to plot the Dow Jones or other financial assets on linear axes. However, for long-range understanding of economic trends, an understanding of exponential growth and the use of log scales to plot data can be invaluable.

Moore's Law

Exponential growth need not necessarily be unwelcome of course. Your view of whether the exponential growth of the stock market was good or bad may depend on whether you had invested in it, and similar exponential growth in house prices has different effects for homeowners and renters. However, there is one classic example of this kind of behaviour where a similar extrapolation has proved remarkably accurate, and where the outcome has been hugely beneficial for all of us.

In the early 1960s, the development of semiconductor technology allowed the creation of integrated circuits in the microelectronics industry, with a particular breakthrough by Robert Noyce at Fairchild Technologies. Integrated circuits allowed tasks which would previously have required many separate silicon chips to be performed by one single chip, giving a breakthrough in performance. It could even be argued that the development of the integrated circuit made the Apollo moon landings possible, with NASA being the largest single purchaser of this technology in the early 1960s.

However, the creation of integrated circuits very much did not rest on these early laurels. The point of this technology was that it could be miniaturised. By shrinking these components and cramming more and more transistors on to a chip, the performance of these chips could grow further. At this scale, the fundamental speed limit becomes the speed of light itself, so that making components smaller allowed faster processors.

By 1965 Gordon Moore, Fairchild's director of research and development, had realised that recent trends in technology indicated exponential growth in the number of transistors on a chip, and hence in computing power. Based on this, Moore made a famous prediction in an article for *Electronics* magazine that:

'The complexity for minimum component costs has increased at a rate of roughly a factor of two per year. Certainly over the short term this rate can be expected to continue, if not to increase.

Over the longer term, the rate of increase is a bit more uncertain, although there is no reason to believe it will not remain nearly constant for at least 10 years.'

In other words, Moore saw that the number of transistors was doubling every year, and extrapolated that this could hold until 1975, predicting exponential growth for the next ten years. Later, Moore revised this prediction to hold until 1980, before suggesting that it might switch to doubling every two years. We can test Moore's prediction by plotting time on the x-axis and the number of transistors on the y-axis using a log scale. Here are the results:

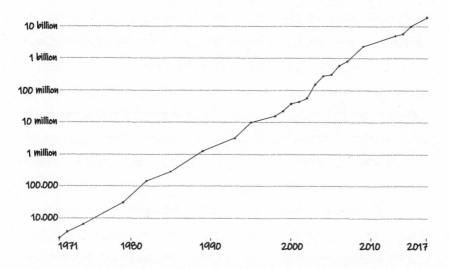

Moore's law: The number of transistors per microprocessor

It isn't a perfect straight line of course. But, by eye, Moore's prediction has clearly worked remarkably well. To all intents and purposes, the data lies on a straight line. But the truly remarkable thing is the x-axis. Time here runs not to 1975 or to 1980, but to 2020. Exponential growth, as tentatively predicted by Gordon Moore in the 1960s, has continued for over 50 years. Chips have grown from under a thousand transistors to over 50 billion.

Indeed, the accuracy of Moore's Law is particularly remarkable given that there is no physical law guiding it, in the way that epidemic spread is driven by the underlying laws of infection. It has simply occurred as the result of continuing innovation at companies such as Fairchild and later Intel, founded by Moore and Noyce in 1968. Many ideas and technological breakthroughs have come together to drive this continuing trend, and despite a variety of confident predictions, Moore's Law has continued to hold, albeit with a small recent slowing, as devices move into the nanometre scale and the sizes of atoms themselves start to matter.

Perhaps the only explanation for the way that actual growth has matched the original prediction is that Moore's Law has acted as a road map. That is, having such an explicit goal articulated by an industry leader has created a target for companies to aim for, and encouraged them to innovate to reach the numbers that Moore's Law would suggest. Similar exponential improvements have been summarised in Swanson's Law, which suggests that the cost of solar panels fall by a factor of four every ten years, showing how technological and manufacturing improvements can provide compound gains.

Whatever the reason, this exponential growth in computing power has shaped the modern world. As our mobile phones have comparable power to a multimillion-pound supercomputer of the past, as we use this to run artificial intelligence algorithms to categorise images or translate language, it is worth remembering the part played by Gordon Moore and his prediction of exponential growth. Indeed, when thinking about issues ranging from pandemics to financial markets, and even the price of footballers, it is important to bear in mind the power of compounded multiplicative growth and of exponentials, and to consider whether using a log scale can bring new insights into the situation.

Summary

In this chapter we have seen examples of exponential growth in settings such as bacterial growth, nuclear reactors, bank interest, the stock market and Moore's Law for the improvement of computers. Of course, it happened at times in the coronavirus pandemic too! I've shown that logarithmic scales and ideas such as doubling time can be useful tools to convert this scary kind of growth into something more comprehensible.

Suggestions

I encourage you to look for examples of exponential growth in everyday life. Most of the time you won't see it, but in the settings that I've mentioned it's worth looking for, because the implications are serious if you do find it. Why not look for websites which plot data and give you the option to switch between linear and logarithmic scales? For example, Our World in Data has coronavirus graphs where you can do this, as do some sites with financial data. By experimenting with these settings, particularly for financial time series over extended periods of time, you can see the effect of switching between these representations. How does the story change? Which way of thinking do you prefer?

Chapter 4
Following the rules

Taking a rain check

Imagine that you are planning a barbecue in two days' time. You've stocked up on charcoal, bought in burgers, buns and beers, and invited your friends to come over. But then you start to worry about the weather and hurriedly check the forecast on your phone. It tells you that there is a 10% chance of rain. But what does that mean, and where does that number come from?

So far, we have seen how a variety of everyday phenomena can be explained by simple mathematical structures, with their behaviour being well described by linear, quadratic or exponential curves. However, this may give a misleading impression of the world – it is certainly not the case that all systems are as predictable as these approximations may imply.

The first thing to say is that although you were quoted a 10% chance of rain, which suggests some element of randomness, the weather itself probably isn't actually a random system at all. The movement of the atmosphere is governed by the laws of physics, so in theory an omniscient being who knew the position and speed of every particle of air measured to infinite precision could calculate the future movement of each particle, taking account of solar radiation, collisions between particles and so on, and know whether there would be rain at your barbecue.

Of course, this isn't feasible for us! We have neither the ability to measure each particle to sufficient precision, nor the computational power to solve the resulting equations if we did. In other words, we will never be able to give a perfect weather forecast. Instead, we do the next best thing and make an approximation.

For example, if we carve the Atlantic Ocean up into 10km squares, and measure the conditions in each one, then we can start to think about how the state of each square might evolve in time. If there's cloud in a particular square, and the wind is blowing at 20km/h eastwards, then we might predict that the cloud would be 20km east, two squares to the right, in an hour's time, then work out where that might go next, and so on. Given a large computer, we can run these calculations forwards into the future and derive a weather forecast for the next few days.

Of course, this will not be perfect. We only receive a snapshot of the conditions in each grid square, and there will be subtle variations within each one. If instead we could obtain weather measurements at a finer scale (say 1km squares) and think about evolution on a five-minute timescale, and if we had an even larger computer to crunch the numbers, then the resulting forecast should be more accurate. The theme of this chapter is carving up space and time into finer and finer chunks and describing how processes follow rules on these smaller scales.

Following the rules

However, the very fact that our weather forecast quotes a 10% chance of rain should give us pause for thought. Just as our omniscient being could solve the equations of motion and get the same answers each time, if we feed the same measurements into the computer ten times then it should give the same answers, because computers follow pre-defined laws of logic and calculation. In other words, we'd expect every forecast to give a 100% or a 0% probability!

In fact, the 10% is an acknowledgement that our operation of carving time and space up is not perfect, and nor are our measurements. As we will see later in the chapter, the equations that dictate the weather can actually behave in a wild and catastrophic way. You may have heard of this idea, first quoted as the title of a paper in 1972: 'Does the flap of a butterfly's wings in Brazil set off a tornado in Texas?' Indeed, slight changes in the measured values can have a huge effect on the outcome of the system.

To get around this, the forecasters acknowledge that the 20km/h wind measurement will be imprecise. To mitigate against this imperfection, they create a collection of forecasts, based around making small tweaks to the measured values – what if it wasn't 20, but 19.5 or 21, 18.7 or 20.5, say? The percentage quoted corresponds to about 10% of these forecasts showing rain at the requisite time.

However, it is certainly worth remembering that computations involving complex systems can be immensely subtle, and it may make sense to think of their outputs as somewhat random. I will say more about the value of understanding randomness in the next part of this book, but for now, we will consider the world of relatively simple processes.

Carving up time

So far, I have described functions changing at fixed time steps. For example, a linear function has the same amount added to its value at each step, and an exponential function is multiplied by the same amount each time. However, of course, time does not move in steps like this. In fact, real-life processes continuously evolve. While we may only learn economic data such as unemployment figures once a month, the underlying state of the economy could change at any day, hour, minute, or even on a finer timescale still. Mathematicians would say that time is a continuous quantity, not a discrete one.

In other words, we should not think of a linear function as a series of distinct points corresponding to separate time values, but rather as a solid straight line, where for any particular time there is a matching value of the function. However, we can still think about the way that the function changes.

We assumed that our space probe was always travelling at the same speed. In a second, we would expect Voyager to travel a certain distance. In a hundredth of a second, we would expect it to travel a hundredth of that distance. For any period, however long or short, the distance travelled divided by the time elapsed would always be the same and would equal the speed.

In fact, we could go the other way. If we were given the speed and the current position of the space probe, we would know where to look for it at a specific time in the future. We could simply multiply the speed by the time elapsed, find the distance travelled, and the future position of the probe would be guaranteed.

It may seem that this argument is based on Voyager having a

constant speed, and to a certain extent it is. However, a version of the argument would work even if the speed were changing over time, so long as we had sufficiently precise information about the speed at every time point. The trick is to assume that although the speed changes, it doesn't change very much over a sufficiently short time. So, for example if we knew the speed now, we could work out how far the probe would move in a second, then find the speed a second later and work out how far it would move in the next second, take the speed a second later, and so on.

This might not give a perfect answer, but it would be fairly accurate. If we wanted a more precise answer, we could repeat the calculation dividing time into hundredths of seconds, or millionths of seconds. Working out the answer might become an involved and annoying sum to perform, but in theory, if we divided time into finer and finer chunks, the answer should get closer and closer to the right one.

What I have just explained is what mathematicians call *integration*. This is a part of calculus, which may be a word that fills you with terror following dimly remembered lessons from school. However, conceptually the idea is simple: given enough data about the space probe's speed, we can find its position. In fact, the other operation of calculus, namely *differentiation*, works in the opposite direction. Given enough data about Voyager's position, we can find its speed.

Pendulums and springs

The key thing is that we can describe the position of the probe in terms of a simple rule involving just the speed. In theory, we could go even further. If we knew Voyager's position and speed at the start, and knew its acceleration at every time, we could do a version of the same calculation. Just as speed tells you how much distance changes, the acceleration tells you how much speed changes. So, we could use a similar process of integration to move from a knowledge of acceleration to a knowledge of speed. Then, we could repeat the steps above to move from a knowledge of speed to a knowledge of position. Hence, using this two-step process, if we knew the acceleration at every time, then we would be able to deduce the position.

Now, and this is where it starts to get mind-boggling, what if the acceleration itself depended on position? This may sound contrived, but this is a situation which arises more frequently than you might think. Isaac Newton's Second Law of Motion tells us that the force applied to an object is equal to its mass times its acceleration. In other words, for an object with fixed mass, the acceleration is proportional to the amount of force imposed on it. But what if the amount of force depends on the object's position?

It turns out that there are natural situations where this occurs. Think of a weight suspended on a spring. Gravity pulls the weight down, and as it gets lower, the spring is lengthened and imposes a larger upwards force, until it exceeds the downwards force of gravity, and the weight is pulled back upwards. Whereas if the weight is higher, the spring exerts less force than gravity and the weight can fall downwards again.

Another example is a pendulum formed by a heavy object hanging on a string. The further the pendulum gets from vertical, the more firmly gravity acts to pull it back towards the centre (essentially gravity always pulls downwards, but the string stops the

Following the rules

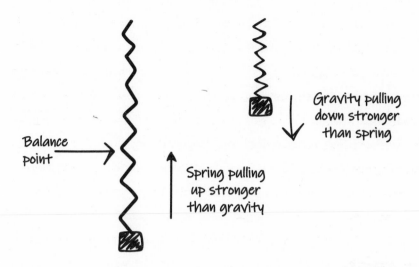

Balance point →

Spring pulling up stronger than gravity

Gravity pulling down stronger than spring

weight from falling vertically, meaning that this downwards force is converted into rotation).

For both the spring and the pendulum, a force acts to push the weight back towards some point of balance. If the weight were stationary at that point, the forces would be exactly equal, and it would remain stationary forever. However, for both the spring and the pendulum, the weight passes through this balance point at a certain speed, and overshoots in one direction or another, before the natural forces reverse the motion again. In theory, in the absence of friction and air resistance, this would go on forever.

In both these cases, the strength of the force is proportional to the distance between the weight and the balance point, and it always acts to push the object back towards the balance point. Hence, because of Newton's Second Law, which tells us that the acceleration is proportional to the force, the acceleration is also proportional to the distance from the balance point, and always acts towards the balance point.

However, the curious thing is as follows: we know from our integration argument that the acceleration determines the position of the object. However, from our description of the forces acting, the position determines the acceleration. It feels like a chicken-and-egg situation.

In fact, the acceleration and position are bound together by these constraints, in a way that determines what type of motion is possible. It turns out that essentially one type of motion is compatible with these rules, namely a kind of oscillatory or resonant (back and forth) movement. This whole description gives a mathematical process known as *simple harmonic motion*, which is exactly the behaviour we see in the movement of a pendulum.

We describe the kind of constraints that give an expression for acceleration or speed in terms of position as a *differential equation*. The operation of finding a trajectory which is compatible with these constraints is referred to as solving the differential equation. This is not always easy, and may require some advanced methods, but in principle it's always possible.

One way to understand the behaviour of this kind of process is via a so-called *phase plot*. For example, here is the phase plot of three trajectories of pendulums, each swung with a different amount of force.

These plots are slightly strange but are worth getting used to. You can see that each trajectory loops round and round, following an oval path. This captures the idea that (ignoring friction and air resistance) an idealised pendulum will swing forever. What is interesting though is what we plot. On the x-axis we plot position (thinking about a pendulum swinging from left to right), on the

Following the rules

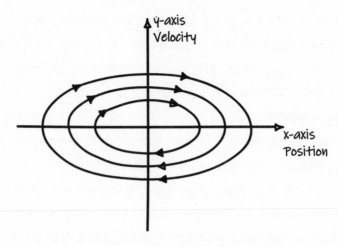

y-axis we plot velocity. The differential equation that governs simple harmonic motion tells us the set of rules that this trajectory must follow: each time the position and velocity are the same, the next steps are the same, and so the pendulum loops round and round on the phase plot, as it swings backwards and forwards in real life. The curious thing is that unlike a standard plot, there is no time axis. Time doesn't matter; we only care about position and velocity.

We can narrate the movement of one pendulum by following the outer loop, say starting at the left-hand side of the diagram. At that point, the pendulum is as far left as it can go and isn't moving (the velocity is zero). Then, it starts to swing back from left to right, speeding up as it goes. As it crosses the y-axis it is back hanging vertically, and travelling at its fastest speed, before it starts to slow until it reaches its farthest point out to the right where again it stops (again the velocity is zero). Having completed this swing, we have moved halfway round the loop. Now, the motion is reversed. We move from right to left, again accelerating and building up velocity, only in the opposite direction to before (below the x axis) before reaching the point where we started narrating and coming to a halt.

This is just one example of a phase plot for one simple system governed by a differential equation. It is possible to produce similar diagrams for more complicated systems, and by studying the trajectory of this abstract representation, mathematicians can gain insight into the evolution of the process.

What is curious though is that it is possible to represent pandemic data in a comparable way. I started doing this in relation to the COVID pandemic shortly before Christmas 2020 – unfortunately, this was a stressful time and so my axes are swapped over compared with the pendulum plots above, but other than that we can think about the following plot in the same way.

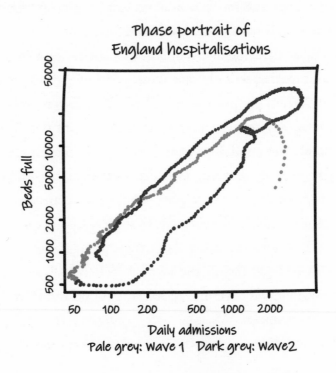

Phase portrait of England hospitalisations

Daily admissions

Pale grey: Wave 1   Dark grey: Wave 2

Following the rules

Here I have represented the first two waves of hospital data in England, and again you will see that the trajectories loop round. I am plotting beds occupied (something like position) on the y-axis and admissions (something like velocity, the rate of change of position) on the x-axis. Again, there is no time axis. The story starts in pale grey midway up on the right, and the points loop round anticlockwise – admissions and beds rising until lockdown, and then a gradual swing round to the left with admissions peaking before beds and then gradually heading down to the left. At the bottom left corner, the colour switches to emphasise the start of the second wave. Again, beds and admissions rise together (apart from a small loop-the-loop caused by lockdown 2) before peaking at a higher level than before, and again swinging down to the bottom left assisted by lockdown and vaccinations.

Indeed, later waves followed similar trajectories, with similar loops taking place. This is one my proudest contributions to the coronavirus data visualisation world. While they are not always straightforward to interpret without explanation, I believe that these plots showed how mathematical ideas of differential equation behaviour could describe how the daily numbers of hospital admissions and beds occupied evolved together. And if nothing else, the pictures look pretty!

Stability and instability, black holes and fractals

Different forms of differential equation can give rise to diverse types of behaviour. As we've seen, one particular choice of equation arising as a mathematical abstraction of problems in physics can lead to an oscillation. But as we know, in real life a pendulum does not swing forever because of friction and air resistance.

We can capture this in our model by adding a small extra term into the differential equation, that gradually reduces the speed of the pendulum, so that it continues to swing, but makes smaller and smaller arcs each time, until it gradually slows down to end up at the balance point. This is an example of a stable system, which moves towards a state of equilibrium. On the phase plot, instead of an oval we would see a spiral towards the centre of the plot.

More dramatic behaviour is possible. Imagine Voyager starting at a balance point out in space but being slowly drawn towards a black hole by gravity. The closer it got to the black hole, the stronger the force of gravity would be, meaning that the acceleration would be greater. Just like our pendulum, the acceleration would depend on position, but instead of the self-correcting motion of the pendulum, the acceleration and speed towards the black hole would become greater and greater with time. In this case, the system would be unstable, in the sense that it would accelerate away from the original balance point.

Obviously, such behaviour could not go on forever because the probe would eventually reach the black hole and be torn apart by gravity, but as a mathematical abstraction it shows that an apparently innocuous change in the differential equation (changing a minus sign to a plus sign, changing from deceleration to acceleration) can have a major impact on the behaviour of the system.

In a comparable way, both exponential growth and exponential decay can arise out of differential equations of a very similar form, again related by just a change in sign. In some sense the differential equation in question is even simpler than those that govern the motion of the pendulum and the accelerating space probe.

We simply consider systems where the speed is proportional to the position. If the speed is a positive multiple of the position, then as the position gets bigger, the speed gets bigger, and the position gets bigger still. This is an unstable system, getting rapidly larger and larger, and exhibiting exponential growth. In contrast, if the speed is a negative multiple of the position, then, like the pendulum, the speed acts to push the position back towards zero. However, unlike the pendulum, the closer the system got towards zero, the smaller the speed would be, and the system would reach a halt exactly at zero instead of overshooting. This is precisely the exponential decay that we discussed previously.

In other words, minor changes in the form of the differential equation can have a big effect on the behaviour of even simple systems. However, much more complicated behaviour can arise from differential equation models, particularly when rather than tracking a single one-dimensional quantity like position along a straight line, we use such equations to model quantities in higher dimensions, such as position in the real, three-dimensional world.

One such notorious set of equations was developed by Edward Lorenz and Ellen Fetter in 1963 to model the behaviour of the atmosphere. I already mentioned that attempting to forecast the weather is an extremely complicated issue, because there are many quantities to model (the conditions at every 1km grid square, for example), so the equations involve many terms.

In contrast, Lorenz and Fetter's equations appear innocuous, in that they have very few terms, and are described only in terms of simple 'speed depending on distance' relationships, rather than needing to consider acceleration. The subtlety in their behaviour arises from the fact that there are three constant values, known as *parameters* to mathematicians, which multiply the various terms.

It turns out that changing the values of these three parameters, even by a small amount, can have a huge effect on the behaviour of Lorenz and Fetter's system. For some values of the parameters, the system just converges to a balance point, in an uninteresting way.

However, other values create much more exciting behaviour. There are certain choices of the parameters which lead the system to not converge to a single point, but rather to be drawn to a strange and beautiful mathematical object called the Lorenz attractor, which is a so-called fractal. This object does not have two or three dimensions, like a sheet of paper or a cup, but rather about 2.06 dimensions. Don't worry if that doesn't make any intuitive sense to you; in fact, join the club. There's no getting around the weirdness of an object having a number of dimensions which is not a whole number!

The fact that Lorenz and Fetter's equations could create such wild behaviour out of simple rules was one of the initial stages in the development of chaos theory. The so-called butterfly effect, referring to the fact that tiny changes in a system can have major impacts, arose as a popular summary of their work. Indeed, a whole area of mathematics, referred to as dynamical systems, continues to study properties of equations like these, and has led to the award of several Fields Medals (essentially Nobel Prizes in maths).

I have described how systems that follow simple rules can behave in strange and unexpected ways. However, for the rest of this chapter I will describe a set of rules that leads to more behaviour which is more predictable but still interesting. Since we know that exponential growth and decay can arise from a differential equation, we can think about how this kind of equation might describe an epidemic and start to understand why epidemics do not grow exponentially forever.

Kermack and McKendrick

The breakthrough work that modelled epidemics using differential equations was published by Kermack and McKendrick in 1927, not long after the Spanish Flu had raged across the world. This was the paper that first introduced the now infamous R number, which they referred to as 'a causal factor which appears to be adequate to account for the magnitude of the frequent epidemics of disease which visit almost every population'.

Their paper introduced what is known as the SIR model, where S, I and R stand for Susceptible, Infectious and Recovered, respectively. The idea is that during a natural epidemic with long-lasting immunity, people naturally move from the first state (Susceptible, meaning they have not yet been infected) to the second state (Infectious, meaning that they may infect others) to the final state (Recovered, meaning that they are now immune from the disease). Of course, this may only approximately describe the true situation, but it remains a useful way to think about the dynamics of disease.

Kermack and McKendrick's work described the rates at which people moved from being Susceptible to Infected to Recovered. However, a key thing to understand is how they moved from thinking about an epidemic as a discrete process (made up of whole numbers of people, at separate time steps) to a continuous one (made up of arbitrary numbers of people).

We can't tell exactly how many people will get infected on a particular day, because that is a matter of random chance. In the same way, if we roll 30 dice per day, we don't know exactly how many will turn up as six on any particular day, but we are confident that the long run average will be close to five.

Just as with Fermi estimation, daily fluctuations on either side of the average are likely to cancel out, meaning that the average gives a good estimate, and so understanding its behaviour is enough – this is sometimes referred to as a *fluid limit*. A result called the Law of Large Numbers (which we will see in Chapter 5) helps give a more formal statement of this.

We can describe Kermack and McKendrick's SIR model through the dynamics of the moves that are made. Susceptible people move into the Infectious category when they are infected. This relies on contacts between the Susceptible and the Infectious subpopulations. If we doubled the size of one subpopulation or the other, the rate of such contacts would double, meaning that the rate of infectious contacts is given by the product of the number of Susceptible and Infectious people, multiplied by some parameter.

Infectious people are assumed to recover at a constant rate, so the number of Recovered people grows by an amount which is proportional to the number of Infectious people. The set of Infectious people grows due to Susceptible people being infected and reduces due to people recovering, but the total population stays constant (we ignore people being born and dying).

These are reasonably intuitive rules that describe how an

epidemic evolves, and lead to three differential equations. These cannot be solved as simply as some of the models described above. However, it *is* possible to find solutions either theoretically or by computer and to understand their properties.

It turns out that early in an epidemic the infected population grows exponentially, and the famous R number (strictly, R0, the initial reproduction number) emerges from the parameters of the model.[6] But this exponential growth phase does not go on forever. Once enough of the population is no longer Susceptible, the rate of growth slows, and the curve starts to flatten. Indeed, one of Kermack and McKendrick's major contributions was to realise that the epidemic would end before the entire population had been infected, meaning that some people would remain in the Susceptible category. In other words, they derived the existence of the famous Herd Immunity Threshold, in terms of the R0 value.

Of course, although Kermack and McKendrick's work was a breakthrough which captures the behaviour of many epidemics, the simple SIR model does not tell the whole story. Various researchers have developed new models to extend it. For example, the SEIR model adds an extra phase, Exposed, where someone has been infected but before they become infectious themselves – for the coronavirus, this phase may last for roughly five days. However, all these models have the common feature that the dynamics of the disease are described in terms of a collection of relatively simple differential equations.

---

6 Specifically, R0 is the rate at which people move from Susceptible to Infectious, divided by the rate at which they move from Infectious to Recovered. This makes sense, because if people are infected faster than they recover then the epidemic will grow (and R0 will be bigger than 1).

However, these equations do not tell the whole story of the spread of coronavirus. For example, in the UK, as we saw in Chapter 3, there were several distinct phases of growth and decline. Similar situations arose in many other countries, with many separate waves due to changes in behaviour and the arrival of COVID variants. This is not behaviour that arises from any conventional SIR or SEIR model. The fact that such models usually only involve one wave of disease led some people to wrongly assume that later waves were not possible, and that the observed decline was due to sufficiently few people remaining in the Susceptible category.

History has not been kind to this way of thinking. In my view, the standard SIR models roughly captured the dynamics of these epidemics once they were adapted in one crucial way. That is, lockdown and social distancing caused a major reduction (at least temporarily) in the rate at which people moved from Susceptible to Infectious. While Kermack and McKendrick's model involved a constant parameter for this rate of movement, a better model would allow this to vary over time (perhaps staying constant within periods of lockdown). In that way, a version of the model based on differential equations would still describe much of the pandemic behaviour observed in the UK.

In any case, it seems clear that understanding such differential equation models is key to knowing how many physical processes evolve over time. So far, we have seen that simple mathematical structures can capture the behaviour of many real-world phenomena. However, real-life data is never as well behaved as these idealisations might suggest. For this reason, we need to discuss the role of randomness, which is the theme of Part 2 of this book.

Following the rules

Summary

In this chapter we have seen how carving up space and time into discrete chunks and describing the rules that a system follows can help explain the behaviour of many processes. Simple rules can describe the motion of well-behaved objects such as springs and pendulums, but also following simple rules can lead to much more pathological behaviour for systems such as the weather. It is striking that Kermack and McKendrick's SIR model for epidemics can be described using a similar set of rules, and that many of the observed features of the coronavirus pandemic, including early exponential growth governed by an R number and the apparent existence of a Herd Immunity Threshold, can be explained by this model.

Suggestions

To further explore the idea of the behaviour of processes being defined by simple rules, you might like to think about the way in which objects move in everyday life. For example, if you go to a playground, you can watch the difference between a child on a swing whose back-and-forth motion follows the pendulum rules and a child on a slide who moves in a consistent direction (though at a varying speed, as gravity and friction play a part). You can think about the forces acting on the tennis ball that cause it to follow the parabola we saw in Chapter 1. Can you find objects such as a lava lamp or a dripping tap whose behaviour, while still governed by the laws of physics, is much more unpredictable?

PART 2:
RANDOMNESS

Chapter 5
Data is random

Coins and Large Numbers

Suppose you have inherited £1,000 and would like to turn that into a larger sum of money. You might invest it in the stock market, hoping to buy shares that are low in price and to sell them when they rise. Alternatively, you might decide to bet on a sporting event: perhaps your football team is playing their local rival, or there's a horse which has an amusing name. Perhaps you'd prefer a game of pure chance, to go to a casino and bet it all on a few spins of the roulette wheel, rolls of the craps dice or draws of cards.

It's fair to say that none of these are guaranteed routes to increasing your wealth, in the way that investing in a fixed rate savings account should be. They all involve some degree of *randomness*, perhaps some more than others. It is hopefully clear that many casino games have an element of pure chance to them and so are inherently unpredictable. However, as we will see in this chapter, there is a convincing case that share prices and sports matches have something of the same character.

For example, consider the stock market, which appears to behave in a completely unpredictable way. On some philosophical level, it may be arguable to what extent the behaviour of share prices is truly random. We might believe that, just like the weather example of the previous chapter, things follow a series of natural laws and that, if only we could model everything in a more precise way, then we could understand the evolution of prices perfectly.

However, we will use the idea of randomness as shorthand to capture the idea that daily share data fluctuate in mysterious ways, some deterministic (stock holdings determined by the needs of tracker funds to match a certain share index, say) and some essentially impossible to predict (did Joe Bloggs decide to buy or sell shares, for obscure and personal reasons?). Certainly, despite the huge financial rewards available for someone who could, it does not appear that anyone can completely predict short-term market changes.

Just as in Chapter 1 we built up an understanding of graphs and curves from simple functions like straight lines, we can start thinking about randomness with a simple object: a coin. If we toss a coin, we have no reason to believe it is more likely to land showing Heads or Tails. We would say that the coin has a 50% probability of showing a Head and will often call this a fair coin.

At other times, it is useful to think about what we call biased coins, which have a different probability of showing Heads, perhaps due to being weighted. In fact, at times in this book we will imagine that we have a complete range of biased coins, so that given any probability – say 83% – we have a biased coin which has that probability of coming up Heads.

Data is random

If we toss any coin several times, past results should not influence the next outcome. As an inanimate object, the coin has no memory, and so there is no mechanism whereby previous tosses can be factored into the result. Mathematicians would say that successive tosses of a fair coin are *uniformly random* (each outcome is equally likely) and *independent* (outcomes do not affect one another).

In contrast, human beings are extremely bad at generating randomness. To illustrate this, I ask you this: Please can you pick a random number between 1 and 100? Most of us believe at a subconscious level that odd numbers 'feel more random' than even numbers, that digits in the middle of the range 'feel more random' than 1s or 9s. For this reason, a disproportionate number of people will pick 37 and 73, and certainly more than half will pick an odd number.

The independence property can cause a certain amount of confusion, because of two very similarly named principles, one a false folklore principle, one a true mathematical fact. The first is the 'Law of Averages'. This suggests that if we have not seen a Head for a while, then the next outcome is more likely to be a Head, to make the sequence appear more evenly balanced. Many people have designed lottery strategies based on this idea, and only choosing balls that have not been drawn lately. Sadly, just like coins, a lottery ball has no memory of past draws and so such strategies are doomed to failure. Arguments based on this idea of a memory are sometimes referred to as the 'gambler's fallacy'.

However, an apparently similar principle, the 'Law of Large Numbers' is true. This tells us that if we keep repeating an experiment with independent results, then the proportion of times that a particular outcome occurs tends to get closer and closer to the true probability of that outcome. This is a subtle property, but the coin experiment illustrates it very well. If we toss a lot of coins, we expect roughly half of them to show Heads. However, we do not expect that this would be exactly the case: if we tossed a million coins, we would be somewhat surprised to see precisely half a million of them give Heads. Equally, we would be extremely surprised to see no Heads at all, though in theory this is a remote possibility.

Mathematicians and statisticians tend to capture this by calculating a range of possibilities that are highly likely to contain the true value. For example, I would ask you to toss a coin ten times and count the number of Heads that you see.

In the table below, I list the probability of seeing each particular outcome, both as an exact fraction and as an approximate percentage of occurrences. You'll see that five (half of ten) is indeed the most likely individual outcome, about two thirds of the time you'll score between four and six. If you managed to get fewer than two Heads or higher than eight, then that's quite a rare outcome (well done!), and the range from two to eight will happen more than 19 times out of 20.

Don't worry too much about how I calculated these probabilities. If you are curious, any particular pre-determined sequence of ten coin-toss outcomes has the same probability of occurring – namely 1/1024. The probability of seeing, say, three Heads is determined by how many such sequences there are which consist of three Heads and seven Tails.

If you have time and a big piece of paper, you might want to check that the numbers 1, 10, 45, 120, . . . crop up as a row of a mathematical object called Pascal's Triangle, formed by writing numbers in a triangular array starting with a 1 at the top of the triangle and with each number being the sum of the two diagonally above it.[7]

Number of heads	Probability (exact)	Percentage of the time
0	1/1024	0.1
1	10/1024	1.0
2	45/1024	4.4
3	120/1024	11.7
4	210/1024	20.5
5	252/1024	24.6
6	210/1024	20.5
7	120/1024	11.7
8	45/1024	4.4
9	10/1024	1.0
10	1/1024	0.1

Now, we'd run out of space to fill in bigger and bigger tables of probability values of this kind. However, using the same ideas, for

---

7 Those numbers crop up here because Pascal's Triangle counts the possible orderings of two types of objects. For example, the third row (if we start counting at zero) contains entries 1,3,3,1. The first 3 corresponds to the fact there are that many sequences with one head and two tails (HTT, THT, TTH). Similarly, there are 120 different sequences of ten tosses formed of three heads and seven tails – or also out of seven heads and three tails, which is why the numbers in the table read the same upwards as downwards.

any fixed number of tosses mathematicians can calculate a range that will contain the true value 19 times out of 20. As the number of tosses grows, this range becomes proportionately smaller, as illustrated by the following table (you can see our previous example of ten tosses as the first row):

Number of coin tosses	Range of Heads	Proportion of Heads
10	2–8	0.2–0.8
100	40–60	0.4–0.6
10,000	4,902–5,098	0.4902–0.5098
1,000,000	499,020–500,980	0.49902–0.50098

For example, with 100 tosses, we can be very confident that between 40% and 60% of them will be Heads, whereas with 1,000,000 tosses this range tightens to 49.9%–50.1%. This illustrates the Law of Large Numbers perfectly.

Expected value and variance

In fact, the Law of Large Numbers holds in many more circumstances than simply describing coin tosses. It brings together two intuitive ideas of 'average' and tells us that if we repeat an experiment enough times these quantities will become close together.

Instead of tossing a coin, suppose we roll a standard six-sided dice many times. (I know that the correct singular form of this word is die, but it sounds wrong somehow.) Again, like the coin tosses, the results of these rolls will be independent and uniformly random. We can note down the outcome of the successive rolls: for example, I just tried this and saw 1, 5, 3, 2, 2, 6, 1, 4, 6, 1 as the first ten rolls, adding up to 31. I would encourage you to find a dice and do the same. Roll it ten times and add up the total of all the rolls.

The *sample average* of the rolls would simply be their sum (31, in my case), divided by the total number of rolls, giving 3.1 for me. In general, a sample average is obtained simply by adding up the numbers observed from repeated runs of an experiment, and dividing them by the number of times the experiment has been performed.

Indeed, we can think of our Head counting experiment in the same way, by writing a 1 on the Head and a 0 on the Tail. The sample average will be the sum of the outcomes, which is simply the number of Heads (each time we see a Head the total goes up by 1, each time we see a Tail the total stays the same) divided by the number of tosses. In other words, the sample average will be precisely the proportion of Heads that we see.

Notice that, because each experiment is random, the sample average is itself random – for the dice experiment we could have in principle seen any value from 1.0 (ten 1s) to 6.0 (ten 6s). The chances are that you won't have seen a total of exactly 31, though of course I can't rule it out! However, it feels as if some values of the sample average are more likely than others. Indeed, we might feel that the value in the middle of this range, namely 3.5, is in some sense the most likely sample average. We might perhaps argue this because 1 and 6 are equally likely, as are the pair 2 and 5 and the pair 3 and 4, with each pair having average value 3.5.

Indeed, we can produce the same kind of table of probabilities as we did for the coin toss example above. The method of calculation is slightly more involved than using Pascal's Triangle, but it produces the following extract from the full table. Note that no particular outcome has a huge probability of happening – the most likely sum (35) only happens about one time in 14. However, there is just over a two-thirds chance that the sum of ten dice will lie between 30 and 40, or that the sample average is between 3 and 4.

Sum of ten dice rolls	Probability (exact)	Percentage of the time
30	2930455/60466176	4.8
31	3393610/60466176	5.6
32	3801535/60466176	6.3
33	4121260/60466176	6.8
34	4325310/60466176	7.2
35	4395456/60466176	7.3
36	4325310/60466176	7.2
37	4121260/60466176	6.8
38	3801535/60466176	6.3
39	3393610/60466176	5.6
40	2930455/60466176	4.8

We can also see this by plotting a graph of the percentage chance of different sums. Clearly there is only a small chance of an extreme value (below 20 or above 50, say – so very well done if you managed to roll values like that!), and that the chances are high that the total will be close to 35.

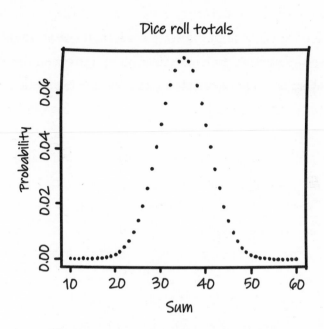

Dice roll totals

The Law of Large Numbers formalises this intuition. It says that if we roll the dice very many times, then it is extremely likely that the resulting sample average is close to 3.5. In fact, this result holds for many kinds of independent experiments. For each experiment, there is a particular number, which we will call the *expected value* (or sometimes the *expectation* or *mean*), associated with it. In general, the Law of Large Numbers says that whatever the repeated experiment is, then the resulting sample average is highly likely to be close to the expected value.

Further, we can describe how to calculate this expected value. Again, it is an average. In the case of a uniformly random experiment, it is the sum of the possible outcomes divided by their number. So, for example, for the dice roll, the expected value is (1 + 2 + 3 + 4 + 5 + 6)/6 or just our 3.5. For the coin toss, the expected value is (0 + 1)/2 or just 0.5.

If the outcomes are not uniformly random, the way we calculate the expected value is slightly more complicated to describe. Again, it is the average of the possible outcomes, but weighted by the probabilities of each outcome. So, for example if the coin is biased, with a probability 2/3 of coming up Heads and a probability 1/3 of coming up Tails, the expected value of the number of Heads would be (2/3 x 1) + (1/3 x 0), or 2/3.

Note that this is a slightly curious sense of the word 'expected'. For example, no dice roll will ever show the outcome 3.5, since we only ever get whole numbers, so there is a sense in which we don't expect to see the expected value at all! We should think of it as a long run average across sufficiently many repeats.

Different experiments can give the same expected value. For example, if we toss a fair coin with a number 3 written on one side and a number 4 on the other, then the expected value would again be (3 + 4)/2 or 3.5, the same as the dice example. However, clearly the values of the dice roll would be more spread out than the values on the coin. The score of each dice roll could be up to 2.5 away from its expected value, whereas this coin would always be within 0.5 (on one side or the other).

We capture this idea that different experiments result in a different spread of outcomes using a quantity called the *variance*. Unfortunately, the details of how we calculate it are slightly too complicated for this book, but in essence, the more spread out the values, the higher the variance. The expected value tells us where we should start looking for outcomes of the experiment; the variance tells us how far out we might have to look to find them.

Expected goals

The idea of expected value has found a new use recently in sports analytics, for example through the idea of so-called Expected Goals (xG) in football. This has been facilitated by access to enormous amounts of data and computer power but has a mathematical idea at its heart. As sports fans, we often want to understand more about the game than simply the final score. Whether our team wins or loses, it can add to the triumph or soften the blow if we know whether the result was a fair one.

For this reason, in addition to the number of goals, many news organisations now report data associated with the match. For example, they may give possession statistics (the proportion of time that each side controlled the ball). It's worth saying that all these metrics can be somewhat problematic and can reveal more about a team's playing style than anything else. For example, a team which seeks to defend deep, absorb attacks and then rapidly counterattack may have a low proportion of the possession, even though this can be a highly effective playing style. Leicester City won the English Premier League title in 2016 despite only having the ball 43% of the time all season.

One traditional way of assessing a football team's performance has been simply to count the number of shots they take. Sometimes this is refined to only count shots on target, but even so this can also be a deceptive measure. That is, an attacking team which is frustrated by a well-marshalled defence may resort to a series of speculative shots from long range which are easy for a goalkeeper to save. Their number of shots would be high, but the chances are that this strategy might not result in many goals.

It should be clear that not all shots are worth the same. A long-range kick from a narrow angle should not count the same as one taken by an unmarked attacker right in front of the goal. To quantify this, data-minded football followers developed the idea of Expected Goals.

Suppose we have a large database of football matches on film and can look at the outcome of each shot. If we group a whole collection of similar shots together, and track the result of each, we can see what such a shot is likely to be worth. For example, if 100 shots from the corner of the penalty area resulted in 12 goals, we might believe that the probability that each such shot results in a goal is 12/100 = 0.12. We can further refine this by considering the position of the nearest defender at the time, for example, separating out shots taken by tightly marked players from those from attackers who find themselves in a lot of space.

This is not an entirely well-defined procedure since it requires human judgement and mathematical modelling, different datasets of previous goals may be used to find the probabilities and so on. For this reason, different organisations may quote slightly different Expected Goals values for the same game. However, the spirit of the calculation is always the same.

Watching the shots that a team takes, we can add together all these fractional bits of goals together, to give their overall number of Expected Goals over the course of a game. This idea has grown out of North American sports – Expected Goals methods were first developed for ice hockey, and the concept of breaking a game down into its constituent parts to measure them lies at the heart of the so-called Moneyball philosophy in baseball. It is no coincidence that such methods are often applied in quantitative finance, where it is possible to trade derivative quantities based on stocks such as the volatility of the market itself, and have often appealed to team owners who made their money in the financial markets.

The reason that it makes sense to add up the fractional values of each shot is really the Law of Large Numbers. If the probability of getting a goal from a particular shot is 0.12 then the expected number of goals that we will see from each such shot is $(0.12 \times 1) + (0.88 \times 0) = 0.12$. Hence, if we take 10 such shots over the course of a game, the expected total number of goals from all these shots will be $10 \times 0.12 = 1.2$ (expected values add up).

However, we should stop and think about what all this means. It's certainly a very clever piece of data-driven analytics work. However, the fact remains that the winner is determined in football by how many goals are actually scored, not by the Expected Goals achieved.

For example, Liverpool fans may have felt hard done by after the 2022 Champions League final, where most xG models gave them a higher value than the actual winners, Real Madrid. However, this is partly due to limitations of the Expected Goals system itself. As described, all that matters is where shots are taken and where defenders are positioned, based on average past outcomes. Of course, this ignores the key factor of the goalkeeper's skill – Liverpool's many shots, even those with high xG values, were kept out by a world-class goalkeeper having a great day. It may not be relevant whether such shots had previously been let in by other less skilful goalkeepers.

There is a further issue though, that of randomness. Just as we don't expect to see exactly 5,000 Heads from 10,000 coin tosses, we don't expect to see the final score being exactly the Expected Goals numbers – not least because these usually won't be whole numbers. In fact, over the course of a single game, there is a fair amount of randomness, and the position may be more like the 'ten coin toss' example.

To illustrate this, consider one of my favourite football results ever. On 4 October 2020, my team Aston Villa played the previous season's league champions Liverpool and triumphed 7-2. This was, to say the least, a surprise based on the two teams' previous form. But even based on the Expected Goals values quoted in the game, it's fair to say the result was somewhat surprising. To illustrate what I mean, I will use the xG values of Aston Villa 3.08 and Liverpool 1.66 quoted by the website understat.com to show that a wide variety of results were possible. The final score did perhaps flatter Aston Villa relative to the xG values reported, but that's partly because some long-distance shots were deflected by defenders, and perhaps other random events happened.

But we can calibrate and illustrate this better by converting the xG values into probabilities, to produce graphs like those that we saw for the coins and dice. The most standard way to do this is to imagine that the actual number of goals is generated randomly by something called a Poisson distribution. It's a standard way of calibrating the likelihood of a large number of rare events – and arises in a variety of settings including, bizarrely, in connection with the

Data is random

number of Prussian cavalry officers kicked to death by their horses. If I feed the 3.08 and 1.66 into the formula for the Poisson distribution, I get the following graphs of probabilities:

Aston Villa goal probabilities

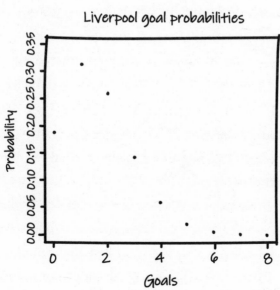

Liverpool goal probabilities

You can see that a wide variety of results were possible. The chances of Aston Villa scoring 7 goals were fairly low – but equally so were the chances of them achieving any particular total (the most likely value was 3, with about a 22% chance of happening). In fact, the model suggests that 40% of the time Villa would have scored 2 goals or fewer from the shots they took, and 23% of the time Liverpool would have scored 3 goals or more.

We can go a step further and imagine that the final goals for each team were drawn independently from the graphs above. This may well not be a realistic assumption, because one team's performance affects the other – if one team leads 1-0 then there is a stronger incentive to attack to get an equaliser; if they lead 4-0 then the opposition may either lose heart or stop attacking to limit further damage.

However, if we assume independence, we can calculate the probability of any given final score. It turns out that the 7-2 result was extremely unlikely. Even with this Expected Goals distribution (representing Aston Villa's exuberant attacking play on the night), we would have seen exactly that result only 0.6% of the time. Under this model Villa would have won 66.1% of the time, 15.5% of the games would have been drawn, and Liverpool would have won 18.3% of the time. In that sense, it's not at all unreasonable to think that the overall result could have gone the other way.

However, I think the real lesson is that part of the glory of sport is its unpredictability, and that attempting to reduce it to a mathematical model can miss the fact that lightning sometimes strikes in unexpected places. At the very least we should not forget that

Data is random

Expected Goals only describes the outcome on average were the game to be played many times. Any one-off fixture has an inherent randomness, which tends to even itself out over the course of a league season, but there will always be a chance of an upset.

Central Limit Theorem and Gaussians

Remember that the Law of Large Numbers tells us that repeating an experiment enough times makes the sample average get close to the expected value. In fact, we can say more than this. For example, since the expected value of the number of Heads is 0.5, the sample average of the number of Heads in 10,000 coin tosses will be close to 0.5. This means that about half the tosses, or about 5,000 of them, will be Heads.

As before, we can calculate the probability of seeing any specific number of Heads, as plotted below.

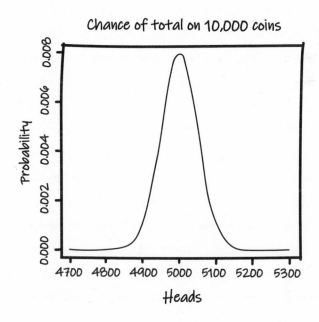

Chance of total on 10,000 coins

We can observe several things. Firstly, no single outcome is particularly likely: the outcome with the largest probability is 5,000 Heads, but even this only has probability equal to 0.8%. This justifies our intuition that seeing exactly 5,000 Heads is very unlikely, despite this number being the expected value. Secondly, we can see that the curve of probabilities has a very characteristic bell shape, sometimes referred to as a *normal* or a *Gaussian curve*. Thirdly, and perhaps most curiously, the curve has more than a passing resemblance to the curve we saw for the probabilities of sums of ten dice. The specific values on the x-axis and y-axis are different, but the shapes of the two curves seem extremely similar.

This is not a coincidence. In fact, normal curves emerge in a universal way from many scenarios involving counting or averaging, so long as the numbers being averaged are independent and are unlikely to take too large values. This means that normal curves are ubiquitous in the modelling of data.

Not only does the Law of Large Numbers tell us that given enough trials the sample average will approach the expected value, but another result (called the *Central Limit Theorem*) tells us that the probabilities of the sample average taking particular values has roughly a normal shape, even for example when tossing a biased coin.

There is a caveat here, which is that although the Central Limit Theorem generally does apply, there are circumstances when it does not. If the items being averaged are not independent or have too large a chance of taking large values (see below for discussion of Extremes), this kind of naive analysis does not apply. This has caused serious problems in the past. For example, many issues of the 2007/08 financial crisis came from assumptions that mortgage defaults in Florida and California were independent, whereas of course they are both driven by the underlying state of the national economy. Hence rises in defaults were more likely to happen simultaneously in both places than the independence assumption might suggest, meaning that total losses could be far bigger than expected from the basic analysis.

Further, it is important to remember that the Central Limit Theorem does not tell us that *all* data fits a bell-shaped curve. Strictly speaking, it only applies in scenarios that arise from summing or averaging large amounts of reasonably independent data. Often these bell-shaped curves do work well to explain datasets that arise in other ways – for example, plotting heights of a random sample of people will tend to show this kind of curve.

However, there are other curves which arise, and relying on intuition based on the Central Limit Theorem can be deceptive, as we shall see in Chapter 7. One reason relates to the symmetry of the bell-shaped curve, where values are equally likely to be the same amount above and below the middle of the curve. In more technical language we might say that the *median* (the value which has half the probability on either side of it) is equal to the sample average.

Not all datasets share this symmetry property, and it can be especially important to realise that reporting the expected value (mean) and median can give a quite different value. For example, this histogram of household income reported by the UK Office for National Statistics (*ONS*) shows that the distribution of incomes is far from symmetric, with the mean being around £7,000 larger than the median. This is because the distribution is skewed, with a relatively small number of high-income people pulling up the mean but not influencing the median to a noticeable extent.

We can use this language to think again about the example of counting emails from Chapter 2. As with the household income data, the existence of some accounts generating extremely high traffic (bots, newsletters and so on) would mean that the median number of messages sent could be considerably below the average.

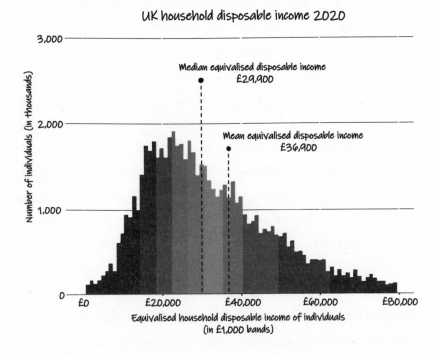

UK household disposable income 2020

However, for incoming messages the distribution is likely to be flatter, suggesting that the median number of messages received could be closer to the average. Another mathematical description is that the variance of the number of messages received might well be smaller than the variance of the number sent, as the values seen might be less spread out.

Data is random

Nonetheless, the Central Limit Theorem is an enormously powerful tool. Not only does it tell us that the result of summing outcomes of repeated experiments will be bell-shaped, but it also tells us precisely what shape to expect. That is, we can think of a range of bell-shaped curves, ranging from a very narrow and tall one to a much wider and flatter one. We could move from one curve to another by stretching or compressing the x-axis (and compressing or stretching the y-axis by a similar amount to compensate).

In terms of our understanding, the narrow bell-shaped curves correspond to scenarios where we can reasonably assume that the value will be close to the expected value, the wide ones where there is more uncertainty. This may remind you of the variance, and this is no coincidence. The Central Limit Theorem tells us precisely that low-variance experiments will lead to concentrated outcomes (narrow curves) and high-variance experiments give more uncertain ones (wide curves).

We can illustrate this by thinking about our two experiments that gave expected value 3.5. Remember that we found that tossing a coin marked with 3 and 4 had the same expected value as rolling a fair dice. However, it turns out that the coin toss has much lower variance than the dice roll (1/4 as opposed to 35/12), so the outcomes are more concentrated.

The Central Limit Theorem tells us what range of outcomes we might see if we performed 10,000 of each type of experiment and summed up the outcomes. We can see two different bell-shaped curves. As we might expect, both are centred on the expected value of 35,000. However, the pale grey curve (representing the coin toss) is narrow, suggesting that nearly all outcomes will lie within 100 or so of that value. In contrast, the dark grey curve (representing the dice roll) is wider, with a reasonable chance of seeing outcomes 300 or more away from the expected value.

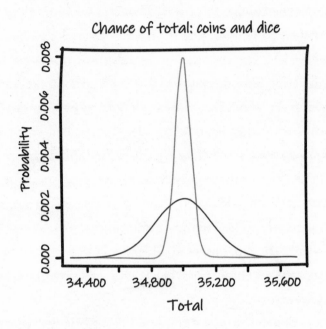

Chance of total: coins and dice

In the same way, the Central Limit Theorem tells us that repeating the same experiment more times leads to a tighter normal curve. This is responsible for the behaviour in the table earlier in this chapter, where multiplying the number of tosses by 100 reduces the width of the plausible range of proportions by a factor of 10.

Extremes

While we have seen that the expected value and variance can be useful ways to capture the centre and spread of random outcomes, this is not always enough. Sometimes we need to understand the so-called tails – that is, to see what extremes are possible or feasible.

Data is random

This is a particular concern, for example, when thinking about the environment and climate change. Weather conditions on 99% of all days may be of relatively little concern to us. However, it is the unusual days which give floods, heatwaves or blizzards that can have a disproportionate effect on our lives. For this reason, modellers often work to capture what might reasonably be regarded as a 'once in hundred years' flood, for example.

Often our intuition is shaky here, because it is based on the common days, not the extremes. A series of factors can combine to produce a value that would have been hard to predict by almost any mathematical model. Think, for example, of Bob Beamon's world record long jump of 8.90m at the Mexico Olympics of 1968. This has only been exceeded once in the subsequent 50 years (by Mike Powell in Tokyo in 1991) but, more than that, it represented a staggering increase on the previous record. Beamon jumped 55cm further than anyone had managed before him, increasing the world record by a greater amount than all previous athletes combined had achieved between 1925 and 1967.

It is hard to think of any statistical model which would consider Beamon's performance as feasible and yet, assisted by altitude and favourable wind conditions, it happened. Similarly, you can probably think of extremes of performance in your favourite sport that are far ahead of the next-best athlete, whether that be Don Bradman's Test cricket average of 99.94, Wilt Chamberlain's 100-point NBA basketball game or Katie Ledecky swimming the 800m freestyle nine seconds faster than any other woman in history.

Of course, these extremes of sporting performance have given us memorable and entertaining moments to watch. However, other types of extreme behaviour can have much more profound consequences. For example, consider the problem of climate change. This requires intensive computer modelling to make formal predictions; however I can illustrate some of the issues by thinking about extreme values. If we hear that, without mitigation, climate models suggest mean global temperatures may increase by 2 degrees Celsius, that may not sound too bad. A pleasant summer day of 23 degrees in the UK might become 25 degrees, which perhaps doesn't seem like cause for alarm.

However, the concern is much more in the extremes. For example, consider the progression of the UK's high temperature record. This has increased by 3.6 degrees in just over a century – starting at 36.7 degrees in 1911, rising in stages to 38.7 degrees in 2019, before a significant jump to 40.3 degrees on 19 July 2022. Indeed, in one day this final jump increased the record by more than the entire amount that average global temperatures have increased so far, believed to be around 1.2 degrees. In other words, to understand the consequences of climate change it is not enough to think about averages, we need to think about the shape of the distribution of temperatures.

If we regard days above 40 degrees as dangerous, the modelled 2-degree average increase would move what might have been a 38-degree day into this range. The problem is that if temperatures follow a normal bell-shaped curve, then previously 38-degree days would have greatly outnumbered the 40-degree days, so the relatively small shift in temperatures could disproportionately increase the number of dangerous days.

Indeed, the picture could be worse than this. If, in addition to increasing the expected value, climate change increased the variance of the distribution, making the numbers more volatile, then extreme values at both ends of the distribution could become more common – even though temperatures might increase on average, there could be an increase in the frequency of both extreme cold and warm weather (though more so at the warm end).

In fact, it is possible that the position could be worse still. I have described scenarios where the range of temperatures has a Gaussian shape. However, there are a variety of curves that are more pathological than this, with a bigger chance of extreme events taking place than a Gaussian might suggest. If you perform calculations of risk based on a Gaussian shape, then this can lead you to severely underestimate the danger, if the data comes from one of these so-called heavy-tailed distributions. This is illustrated in the following figure, taken from a 2012 IPCC special report, 'Managing the Risks of Extreme Events and Disasters to Advance Climate Change Adaptation'.

Indeed, faulty Gaussian assumptions already led us astray during the financial crisis of 2007/08, when the standard models of financial markets which I will describe in Chapter 10 led to a serious miscalibration of risk. The basic versions of such models assume that fluctuations in stock prices have a Gaussian distribution, yet there were daily swings that were far greater than these models would predict.

The CFO of Goldman Sachs was quoted in the *Financial Times* in August 2007 as saying, 'We were seeing things that were 25 standard deviation moves, several days in a row.' This may sound technical, but I can illustrate what this means. From a Gaussian,

Numbercrunch

Shifted Mean

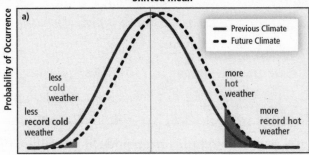

Increased Variability

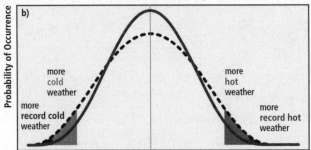

Changed Shape

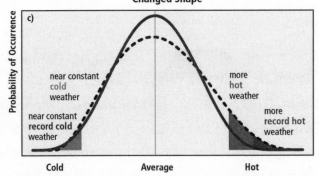

The effect of changes in temperature distribution on extremes. Different changes in temperature distributions between present and future climate and their effects on extreme values of the distributions:
(a) effects of a simple shift of the entire distribution toward a warmer climate;
(b) effects of an increase in temperature variability with no shift in the mean;
(c) effects of an altered shape of the distribution, in this example a change in asymmetry towards the hotter part of the distribution.

we expect to see a value more than 2 standard deviations (another measure of spread, in fact the square root of the variance) away from the expected value about 5% of the time. Because of the bell-shaped curve, the probabilities of bigger values drop away extremely fast, and the chance of seeing a value 25 standard deviations away from the expected value is about 10-136, a ludicrously tiny number. We never expect to see this happen during the lifetime of the universe, let alone several days in a row. What this tells us is simply that Goldman Sachs' models were wrong and underestimated the chances of extreme events with disastrous consequences for the world economy. For this reason, it is important to understand the extreme properties of a dataset, as well as its centre and spread. Making standard but incorrect modelling assumptions lulled us into a false sense of security about the extreme possibilities of how bad a financial market crash could be, and the same could be true for climate change.

Summary

In this chapter, we have developed language to characterise randomness, to compare different random processes and describe the features of data. This has included ideas such as the expected value and variance, which capture where the data is centred and the size of the fluctuations you might see around that value. Further, the Law of Large Numbers and the Central Limit Theorem tell us about the long-run average of the process, for example showing how the total fraction of Heads tends to become closer to 1/2 the more tosses we do. We have seen how this language allows us to understand metrics such as the Expected Goals in a football match, and how also thinking about extremes allows us to build an understanding of the potential for devastating consequences of climate change or financial contagion.

Suggestions

You might like to try out some of the simple experiments I have described with dice and coins. It would be interesting to see how the results you obtain compare with the theoretical distributions I have described. You might like to think about other random processes. If you play Monopoly, what would the expected total score of rolling two dice together be? What are the chances of rolling three doubles in a row and so having to 'Go to Jail'? If you buy a lottery ticket, what are your expected winnings? What is the most likely amount that you win? How would the answers to these questions change if, for example, there is a rollover and last week's jackpot is added to this week's prize fund?

Chapter 6
Vital statistics

Statistics and hypothesis tests

Imagine that a pharmaceutical company has developed a drug which might help prevent strokes. It's expensive to produce, and it's possible there might be some mild but unpleasant side effects, so we don't want to recommend the drug for the sake of it. On the other hand, if it does work, this could be a game changer! How do we decide whether to license the drug?

The key idea is that we should carry out a clinical trial. There are many subtleties to the way that this is usually done, involving two groups of similar people, one set receiving the drug and the other receiving a harmless placebo, with patients assigned at random to each group and nobody, not even the doctors carrying out the trial, knowing who receives which treatment.

However, I will describe a simpler version of this idea. Imagine that at present 1% of 70-year-old men have a stroke every year. (Don't worry about this figure, it's one I just made up!) Suppose that the pharmaceutical company recruited 1,000 randomly chosen men of this age into a trial and gave them the drug for a year. At the end of this time, they found that only 5 of them have had strokes. Should we license the drug?

On one hand, this seems great! In the language of Chapter 5, if the drug has no effect then the expected value of the number of strokes in this group is 1,000 x 0.01, or 10. Clearly 5 is less than 10, so it seems like the drug works! On the other hand, also from the previous chapter we know that strange things (Aston Villa scoring 7 goals against the league champions) can happen by random chance. Could this be one of them?

Rather than making decisions based on instinct, there is a principled way to decide, based on probability and statistics, and specifically the key idea of a *null hypothesis*. In general, a null hypothesis refers to a kind of default belief about the world – for example, we might start from a prior belief that the new drug has no effect (either positive or negative) compared with the status quo. Since changing our view of the world and adopting this new treatment requires money, work and effort, we will only do it given a certain amount of evidence, specifically a trial result which is 'not the kind of thing you'd usually see from the standard treatment'. So, the key question is: how surprised would you be to see 5 strokes in a year in this group if the drug had no effect?

We can understand this better by returning to the analysis of the fair coin from Chapter 5. The idea is that we can use the table of plausible outcomes from that chapter as a guide to decide if a coin might be fair. We've already seen that the number of tosses makes a difference, and in the same way the size of the sample matters.

Vital statistics

It may well be that we'd make a different decision if we'd seen 500 strokes from a random group of 100,000 men.

If we toss a coin 10,000 times and see 5,072 Heads then, based on the table from Chapter 5, this is within the range of results that we might expect to see from a fair coin, and so gives us no reason to disbelieve the (null) hypothesis that the coin is fair. If the same experiment gives 5,200 Heads, this is not the kind of outcome that we tend to see from a fair coin, and therefore we should reject the hypothesis of fairness. This result of 5,200 Heads is referred to as *statistically significant* – this is an important status to achieve, and this terminology should only really be bestowed by a statistician, not based on informal eyeballing of the result!

In fact, we can be more precise. It is possible to calculate that, if the coin is fair, being 200 or more Heads away from 5,000 is the kind of result that we would only see once in 16,505 experiments. On this basis, it seems extraordinary to believe that such a result has happened at random, and so we have convincing evidence that the coin is not fair.

This 1 in 16,505 chance, or a 0.006% probability, would be referred to as a *p value*. Formally speaking, this is defined as the probability of seeing a result at least as extreme as the data, assuming our null hypothesis is true. By convention, many people and scientific journals regard any p value smaller than 5% as sufficient evidence to reject the hypothesis. Clearly, this is a somewhat arbitrary threshold, and in practice it may simply be best to quote the p value directly. But essentially, the smaller the p value, the less likely it is that this result happened by accident, and the more justifiable it is to reject the null hypothesis.

Conveniently, since our table lists a range of values which occur with probability 95% (assuming the coin is fair), it gives us a way to decide whether the coin is fair. For example, if we toss the coin 100 times and see a total number of Heads outside the range from 40 to 60, then we could feel justified in rejecting the hypothesis that the coin is fair.

Returning to our stroke trial, we can do a similar calculation and find that you'd see 5 or fewer strokes by pure chance in a sample of 1,000 people about 6.6% of the time. So, the result is suggestive, and feels like it merits further investigation using a larger trial. But it wouldn't be statistically significant, at least at the 5% threshold which is standard in the scientific literature.

However, a word of caution is necessary. Think about a thousand people separately tossing fair coins 100 times each. Roughly 19 out of 20 of these people would see between 40 and 60 Heads. However, 1 in 20, or around 50 people in total, would see a total number of Heads outside that range. Each of those 50 people would therefore wrongly believe that their coin is not fair. In the context of clinical trials, this could lead us to believe that 50 treatments which have no effect were in fact effective – luckily, statisticians have tools to deal with this by requiring stronger evidence when more potential treatments have been tested!

However, as the xckd webcomic overleaf parodies, there is a danger of misleading results being published by accident by this route. For example, if a thousand scientists independently set out to test the existence of ESP, and if only the 50 whose results are statistically significant send them for publication, then from the published literature it might appear that ESP did indeed exist. As a result,

Vital statistics

clinical trials should be registered in advance to avoid this so-called file drawer problem (sometimes also referred to as publication bias).

Think of your friend with the glamorous Instagram profile. Judging from the photos they post, they are always having a great time – drinking fancy cocktails, eating in expensive restaurants and watching the sun set from sandy beaches. But you need to remember that they don't share photos of the times when they are at home and eating a six pack of reduced-price sausage rolls in one sitting. Just as your friend's carefully curated but unrepresentative collection of Instagram photos gives a deceptive impression of their life, researchers only choosing to publish a selection of their experimental results can be misleading about the validity of the effects they have observed.

Similarly, either through unscrupulousness or misunderstanding of statistical principles, it is possible for one researcher to subject the same dataset to very many different statistical tests, looking for a hypothesis that it satisfies by random chance even when a statistically significant effect is not present. This is sometimes referred to as p-hacking and is again a potential problem in the scientific literature.

These issues mean that statistical procedures are not an infallible way to decide things. Indeed, a certain amount of error is baked into them, because apparently extreme events do sometimes happen by pure random chance. It is important to remember this when results of trials are discussed.

Confidence intervals

We can use reasoning based on the Central Limit Theorem not just to test whether a hypothesis seems to be true, but also to estimate quantities of interest and give margins of error on those estimates.

Vital statistics

For example, suppose we want to measure the chance that a randomly chosen British person likes gherkins – call this probability p, and remember that this is not known to us. In theory, we could carry out a census: survey every person in the country and ask each one whether they like gherkins. However, this would be expensive and time-consuming: a more practical way to find p would be to carry out an opinion poll. Instead of asking everyone, we could take a large, randomly chosen representative sample of the UK population and ask each of them the same question.

There are some technical issues about how we carry out such a poll, since finding a representative sample is easier said than done, an issue that I will return to in Chapter 11. However, we will imagine that these sampling issues have been resolved, in which case the Law of Large Numbers tells us that the proportion of gherkin lovers in the sample should be close to the true probability p.

From an alternative perspective, we could say that the (unknown) probability p will be close to the proportion of gherkin lovers in our sample, that we have discovered from the survey. In fact, in the absence of any other information, the most sensible guess of p would simply be this proportion. Statisticians call this a *point estimate* of p, partly because 'estimating' sounds much more scientific than 'guessing'.

In general, it is good practice to not just quote the point estimate of p. Just as we do not expect to see exactly 5,000 Heads when tossing 10,000 fair coins, we do not expect that the proportion of gherkin lovers in our sample will be exactly p, so the point estimate is unlikely to be perfect. In fact, we can understand this uncertainty using an argument based on the Central Limit Theorem.

Just as in the table in Chapter 5, where the calculation was performed for fair coins (p = 50%), for any given p, statisticians can use the Central Limit Theorem to work out a range of values of the proportion of gherkin lovers that we might expect to see (this is represented by the forward arrow marked 'CLT') on the diagram below. Again, we can use the alternative perspective and ask which values of p are compatible with the observed proportion in this sense (this is represented by the backward arrow marked 'CI').

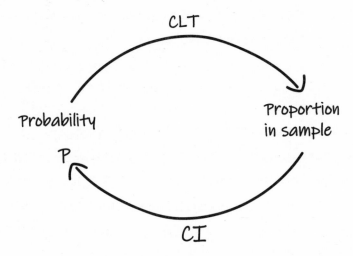

This process gives a range of possible values of p that we reasonably believe might be true: this is called a *confidence interval*. To be precise, because we looked at a range of proportions that would contain the right answer 19 times out of 20, strictly speaking, this is a referred to as a 95% confidence interval.

This is illustrated in the figure overleaf, which is based on interviewing 100 people (the more people interviewed, the thinner the resulting lozenge shape). For any particular p, we can calculate

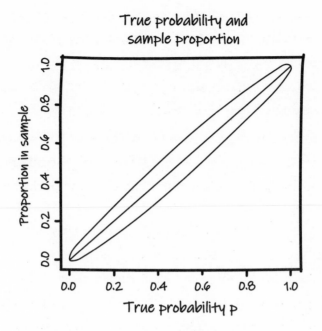

**True probability and
sample proportion**

the range of proportions we might expect to see, as in the table
from Chapter 5, and plot it as a vertical interval of values positioned above the value p on the x-axis. Collecting these intervals
from left to right, we can build up the lozenge shape, which represents the true probabilities and sample proportions which feasibly
go together. The lozenge will contain a relatively narrow region
either side of the diagonal.

Now we can go the other way. Given a sample containing 45%
gherkin lovers, we can draw a line horizontally from 0.45. The
region of the lozenge that this line intersects is the set of values of
p that are compatible with the observed 45% figure. Roughly speaking, this will be a small interval either side of 45% – running from
35% to 55%, say. That is our confidence interval for the proportion
of gherkin lovers in the UK population, based on a sample of 100.

It is always good practice to provide confidence intervals rather than just point estimates, to show how uncertain the estimate is. In UK politics, it is standard to take an opinion poll of 1,000 people – this is often quoted with a 'margin of error' of 3%, which is a confidence interval exactly as described above. In general, the larger the sample polled, the narrower the confidence interval.

To calculate the confidence interval can be an involved procedure, requiring an expert statistician. I certainly don't expect you to be able to perform the calculations yourself. However, any properly presented scientific paper will contain not just estimates of quantities, but the confidence intervals around them.

In an ideal world, these confidence intervals would be reported by the media – so the headline '45% of UK population like gherkins, new study finds' should be followed by some discussion of the width of the confidence interval around this, so that the reader can judge how much credence to put in any quoted value. At the very least, a news story or university press release should link to the original research paper, so that the interested reader can look up the confidence interval for themselves.

As a rule of thumb, although the point estimate is in some sense the most likely value, any number within the confidence interval should be regarded as plausible. If we had previous evidence, we would not typically abandon that if the corresponding value lay in the confidence interval. For example, if previous surveys showed that 40% of people consistently liked gherkins, then based on our 35% to 55% confidence interval we may well decide that nothing had changed, and that this past figure still held true.

Regression

Another statistical calculation which is often extremely useful is known as *linear regression*. This explores the relationship between two types of data and helps us understand if there is a correlation between them, that is whether increasing one piece of data tends to systematically increase (or decrease) the other.

For example, suppose we believe that eating pies causes weight gain. We could take a sample of the population, weigh them and record how many pies they eat in a week. We then plot the resulting data on a graph: the number of pies lies on the x-axis, and the weight (in kilograms) on the y-axis. By convention, the quantity that we are trying to explain should always appear on the y-axis, the quantity we use to explain it should appear on the x-axis. Each person interviewed corresponds to a point in space.

We would like to find a relationship between the number of pies and the weight. One way to do this would be to imagine that there is roughly a linear relationship: each pie eaten weekly typically adds a certain amount to someone's weight. We could test this by trying to find a straight line going as close to the points as possible, which we call a *best-fit line*.

We would be extremely suspicious if the points all lay exactly on a straight line – as with the overfitting example of Chapter 1 this would seem too good to be true – but we might believe that they could be close and will think of departures from the line being explicable by random fluctuations.

This is illustrated here. There seems to be reasonably convincing evidence that the theory might be true. The best-fit line goes

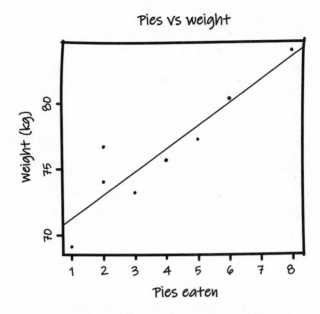

reasonably close to all the points, and has a strong upwards trend, suggesting that the more pies eaten, the greater the typical weight.

The details of how this best-fit line is calculated are too technical to discuss here, but the line should go through the point that corresponds to the two averages (that is, a point whose x coordinate is the average pie consumption, and whose y coordinate is the average weight). Having done that, we can look for the best possible slope, tilting the line up or down to get it as close to as many points as possible. Since we believe that the points are somewhat arbitrarily placed due to the effect of random noise, we should not regard this line as definitive. In fact, it is reasonable to think that a range of values of slope are roughly consistent with this data, and we should ideally give a confidence interval for the slope.

We can return to the iced bun example of Chapter 1 to illustrate this. Remember that although our data appeared to show a possible rising trend, we could be led into a misleading feeling of confidence by drawing a polynomial curve through all the data points, which appeared to make crazy predictions for future weeks. In contrast, a linear fit seems much more believable. The best-fit line doesn't go exactly through all the points, but it seems close to all of them.

Further, the predictions seem much more modest: the slope of the line suggests an increase of two buns per week, which feels like a more reasonable and sustainable trend to extrapolate than the wild polynomial predictions. Indeed, we can calculate a confidence interval for this slope, which runs from 0.1 to 3.7. In other words, we can't rule out the possibility that there is essentially no increase at all in sales going on.

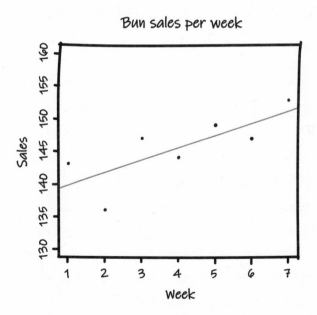

Although in the pie and buns examples the lines seem to fit the data well, there are some caveats to be considered before jumping to conclusions. Firstly, these are small samples, of only eight people and of only seven weeks respectively. We would like to see more evidence than that before deciding for sure.

Secondly, you may have heard the phrase 'correlation is not causation'. All that these graphs illustrate is a correlation – when one quantity is large, so is the other. In theory, we could swap the pie graph around, plot weight on the x-axis and pies on the y-axis. The line would be equally close to all the points, but would we feel justified in concluding that having a larger weight causes someone to eat more pies?

Thirdly, there may be some common factor that causes both variables to appear related in this way. Perhaps old people tend to weigh more than the young, and also old people have a love of pastry. In an ideal world, rather than simply presenting a correlation of this kind, we would attempt to control for factors like age, socio-economic status, gender and so on.

There can also be a temptation to present graphs with a best-fit line plotted that implies some degree of correlation, when in fact the evidence is very weak. For example, if we asked the same people how many books they had read in the last month, plotted that, and found the line that came closest to the points, we may produce a graph like the one on the next page.

Again, there appears to be a trend: the line slopes up, suggesting that the more books read, the greater the weight. However, it is worth noticing that the line isn't particularly close to many of the points, suggesting that the correlation is weak.

The strength of correlation is often measured using a quantity called R^2, which always lies between 0 and 1, and captures

Vital statistics

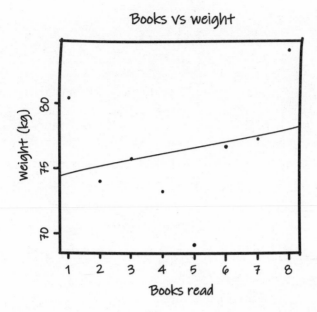

Books vs weight

how close the line is to the average point. The further R^2 is away from zero, the more convincing the correlation. For example, the pies example has R^2 of 0.84 and the iced bun example has an R^2 of 0.61, which confirms that there seems a good correlation. In contrast the books example has R^2 of 0.07, showing that any relationship is very weak.

Another way to see the weak correlation is to consider the flatness of the line. In an extreme case, if the line was completely flat, then the number of books read would have no effect at all. In this case, from a statistical viewpoint we cannot rule out the possibility that the true line is flat. We should look for a p value for the hypothesis that the slope is zero – that is, a probability that we'd see a fit that good if there was no effect at all. We can summarise the results in each case as follows:

Numbercrunch

Example	R²	p value	interpretation
Pies	0.84	0.14%	strong correlation
Buns	0.61	3.91%	strongish correlation
Books	0.07	53%	likely no correlation

However, note that the specific interpretation in the right-hand column is a matter of degree. While smaller R^2 and larger p values are evidence of weaker correlation, there is no specific threshold for either that implies there is definitely no correlation.

Predictions and interpolations

One crucial point to consider is the difference between modelling *current* data and predicting *future* data. In general, the first is relatively easy but of limited utility, but the second is considerably harder and much more valuable.

Given a range of data points and a best-fit line, we can be reasonably confident about interpolating within the range of values that we have already seen. Given our data concerning people who have eaten between one and eight pies, then if we learn that someone eats seven pies then we can try to estimate their weight based on that.

The natural first idea is to simply estimate their weight using the corresponding value on the best-fit line. However, this is a mathematical abstraction, and as we know, not all the points lie exactly on the line. Instead, we should be looking for a confidence interval which has a good chance of containing the true value.

As I have described, the position of all the points is somewhat random, and the weight of this new person will be no different. This will make the confidence interval around our prediction be slightly wider than you might expect, because we must combine the uncertainty about how close to the line the new value will be with our uncertainty about the slope of the line. However, we can use this to estimate the weight of the seven-pie-eating person, giving a range of potential values centred on the value given by the best-fit line, but with a calibrated amount of uncertainty around it.

However, if we wished to estimate the weight of someone who ate 100 pies, there would be two problems. First, as I have described, the estimate of the slope is uncertain. This has a relatively small effect when considering points close to the centre of the range of pie consumption values for which we have data. However, for points much further away, the uncertainty of the slope has a bigger effect, since small pivots in the angle of the line are magnified by the effect of distance (think of trying to adjust the position of a very long ladder, standing at one end of it). In any case, it may be sensible to think whether we even believe that a linear relationship should hold over such a wide range of values.

There is a second issue when seeking to extrapolate in this way, which is that we have assumed the randomness for each point was generated in the same way. However, it might be reasonable to ask if this is the case: having based our model on the study of people who eat between one and eight pies, can we really be confident that the same factors apply to someone extremely far outside this range?

One more cautionary message regarding regression is that not all data is well explained by a linear relationship, and that properties of a dataset can give a deceptive picture. This was captured beautifully by a collection of datasets known as Anscombe's Quartet, named after the statistician Frank Anscombe, who introduced them in 1973.

The key is that regression calculations could in theory be done without even looking at the dataset. That is, the line that we plot will always go through the point that corresponds to the two averages, and its slope will be determined by properties of the dataset such as the variances and correlation, which we refer to as summary statistics. The point is that different datasets can have the same summary statistics, and hence the same line. However, in some cases this line captures the features of the data much better than others.

This is illustrated by Anscombe's Quartet, which consists of four datasets, each with the same summary statistics, and hence the same best-fit line. The first dataset is one of the kind we have been looking at, where the line does a reasonable job of explaining noisy data. The second one is not well explained by a straight line – clearly a parabola (like the tennis ball throw of Chapter 1) would do a much better job. The third dataset has a clear outlier. That is, all but one of the points lie on a different straight line, but that point is quite different to the others and distorts the choice of straight line used. In practice, we would hope to spot such an outlying point and investigate it further, to see if there is a data recording error or another reason to believe it might behave differently. The fourth of Anscombe's datasets is another example where a linear relationship may not be appropriate. Nearly all the points have the same x value, but very variable values of y, and the angle and position of the line is driven by the point very far out to the right, which again could be an outlier for some reason, and to which it might not be reasonable to overfit.

Anscombe's quartet

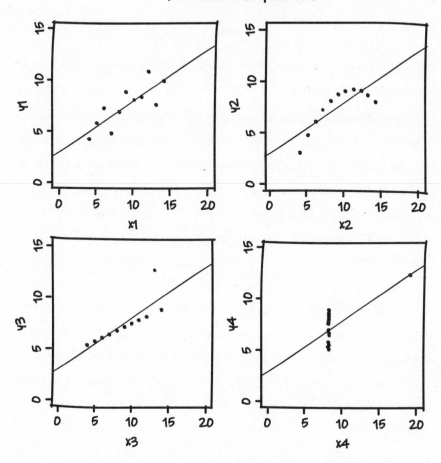

The curious thing about Anscombe's Quartet is that each of these fitted lines has the same R^2 value, implying that the quality of the fitted line is equally good in each case. This is perhaps true, but equally it seems clear that other functions could do a better job, such as the parabola in the second dataset and the other straight line in the third dataset should the outlying point be removed. For this reason, the lesson of Anscombe's Quartet is that we shouldn't just rely on mechanistic procedures to fit regression lines by computer, but should rather examine the dataset carefully first, for example by plotting it.

Statistics and the pandemic: ONS, REACT and North West England hospital beds

Having seen all this, we are now in a better position to see how statistics can help make sense of pandemics, with reference to COVID in the UK. Incidentally, as an aside, the word 'statistics' is often used interchangeably with the word 'data'. Personally, I try to avoid this: I tend to use data to refer to raw numbers being announced, and statistics to refer to the by-product of analysis performed on these numbers. However, I accept that this is a losing linguistic battle.

One place where statistics played a vital role was in the weekly Infection Survey reported by the UK's Office for National Statistics (ONS). This worked roughly like the survey of gherkin preferences described above. Each week a large group of people (roughly 100,000) were tested for coronavirus. By counting the positive tests, the ONS could estimate the *prevalence* of COVID (that is, the percentage of people who have the disease across the country). Further, by splitting the large sample into smaller subgroups by geographical area or by age, the ONS could also estimate the prevalence by region or within age groups. Finally, since the results were recorded each week, it was possible to track trends – had the total infected population gone up or down and by how much?

This was a vital contribution to our understanding of the spread of the pandemic. Since the group was sampled randomly, we could be confident that the figures were representative – in contrast to case numbers, which depended considerably on the availability of tests and whether testing targeted virus hotspots. In that sense, these ONS results were the gold standard for estimating prevalence and were widely reported by the media.

However, while the ONS were careful to report their estimates of the prevalence including a confidence interval, this caveat did not always feature in the press coverage. As I have explained, a confidence interval is necessary in understanding trends, since the true figure may be plausibly anywhere within the quoted range.

You may be interested to see some of these point estimates and confidence intervals. For example, between 8 and 21 June 2020, the ONS estimated 0.09% of the population of England were infected, but quoted a confidence interval from 0.04% to 0.19%, or between 22,000 and 104,500 – almost a factor of 5 between lower and upper bounds. Roughly speaking, at that prevalence, with a sample of 100,000, we might expect to see only 90 positive tests – so small random fluctuations (one or two people performing the test wrongly for example) could play a major role in changing the estimate.

In contrast, between 27 March and 2 April 2022, the ONS estimate of prevalence had risen to 7.60%. This was contained within a confidence interval from 7.40% to 7.79%, or between 4,070,000 and 4,284,500 – a much narrower range, proportionately speaking. At that higher prevalence, our sample might contain around 7,600 positive tests, so a few badly performed tests would have less effect.

A more sophisticated analysis was performed by the Real-time Assessment of Community Transmission (REACT) survey, based at Imperial College. Again, this was carried out somewhat like an opinion poll, sampling around 10,000 people per day throughout a two- or three-week period. By counting positive tests day by day, REACT could estimate the daily prevalence, and hence deduce the daily trend in infections, allowing it to estimate the R number, to tell how fast the pandemic was increasing or decreasing in size.

This was again a very impressive piece of analysis and came with confidence intervals on the estimate of the R number. However, again these were not always reported as they should be. This could lead to curious, outlying values being quoted, without any attendant impression of the uncertainty around them, which is far from ideal.

The effect was exacerbated when the sample was split into regional subgroups, because this led to smaller daily sample sizes and hence to greater uncertainty. For example, at the end of October 2020 many newspapers reported that REACT's Interim Round 6 results had estimated an R number of 2.06 in the South West – suggesting the alarming possibility that the epidemic might be doubling every five or six days. However, they were less likely to report the confidence interval, which ranged from 0.98 (epidemic shrinking) to 3.79 (growing faster than in March). Given the width of this confidence interval, it was hard to feel that the survey could say anything definitive about this subgroup, and certainly the quoted headline value of 2.06 required caution.

Indeed, this scepticism was somewhat justified by the remainder of Round 6, with a final estimate published two weeks later which estimated an R number of 0.95 (confidence interval from 0.51 to 1.53). Since extremely large-scale changes in behaviour are required to change R from over 2 to under 1, this suggested that a plausible explanation is that the first true value was somewhat lower than 2.06, and the second true value perhaps somewhat larger than 0.95. Certainly, quoting these results with the confidence interval gives context to judge whether such an explanation is consistent with the data.

Having described how confidence intervals naturally arose in the analysis of coronavirus numbers, I will give a personal example of how linear regression can help in this context. This concerns the number of COVID patients in hospital in the North West of England in early September 2020, and I will explain why these numbers alarmed me. All this data is taken from the invaluable PHE (later UKHSA) dashboard.

In the first wave of the virus, the number of beds occupied by COVID patients in the North West NHS region peaked at 2,890 on 13 April 2020. Over the summer, as lockdown measures continued to work and patients were discharged, these numbers fell steadily, showing roughly exponential decay. By 26 August 2020, the number of patients had fallen to 77, which represented a tiny fraction of the peak value, and which turned out to be its low point.

However, in successive days after this, there was a slow but steady rise, with successive days' figures being: 77, 85, 87, 103, 102, 113, 117, 112, 124, 130, 133, 139, 164, 166, 173.

The question was how alarming this trend was. On one hand, the total additional increase wasn't huge – 96 more beds occupied in 14 days, or an increase of seven beds per day. At that rate, it would take over a year to reach the previous peak, which might have suggested that there was little to be worried about.

However, I was immediately concerned, because this analysis was based on linear growth (growing like the distance Voyager had travelled), whereas as I described in Chapter 3, the natural model for epidemic spread is exponential growth or decay (growing or shrinking like a bank account). On that basis, things seemed much more worrying – numbers had roughly doubled in 12 days. In that sense, to reverse the previous decline required occupancy to grow by a factor of 37, or just over five doublings, suggesting that the previous peak might be exceeded in two months or so.

However, since figures on individual days should be treated as somewhat random, it would be foolish to simply draw a line between two points and extrapolate from there. A better solution is to carry out linear regression, using the natural model of exponential behaviour to motivate plotting this data on a log scale, and looking for a best-fit line in this representation.

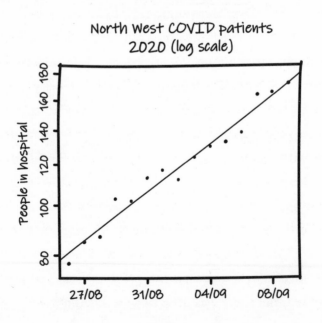

North West COVID patients
2020 (log scale)

Vital statistics

The results of this preliminary analysis are shown here. While the points do not lie exactly on the straight line, there seems to be a strong enough correlation that the possibility of exponential growth should be taken seriously.

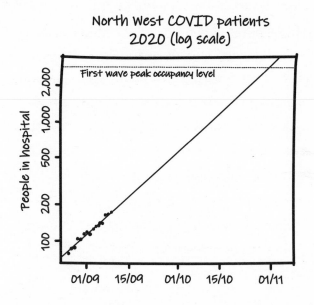

North West COVID patients
2020 (log scale)

A more worrying version of the same graph is plotted here. This seems to confirm the analysis that uninterrupted exponential growth at this rate would lead to the first wave peak being exceeded at the end of October 2020. At that stage, although I warned about it on Twitter and in the *Spectator*, the level of 3,000 or so occupied beds seemed extremely remote. As I have described, there are caveats in extrapolating linear trends for long time periods, since people modify their behaviour, hospital admission criteria change, or restrictions are brought in to combat the spread of the virus.

In fact, the trend, or something like it, continued for an extended period, and although there was some flattening of the curve in late October, the previous level was exceeded on 9 November.

It is possible that this was a lucky guess on my part. There are many famous examples where predicted exponential growth has not come true – whether it be Thomas Malthus's models of population or *The Simpsons* character Disco Stu's optimistic extrapolations of the future popularity of disco music. However, given the severe risks of exponential growth and the fact that, as described previously in Chapter 3, it is the default model for the way that a pandemic should behave, it seems prudent to take the possibility seriously, and using linear regression of data on a log scale plot remains the easiest way to investigate such claims.

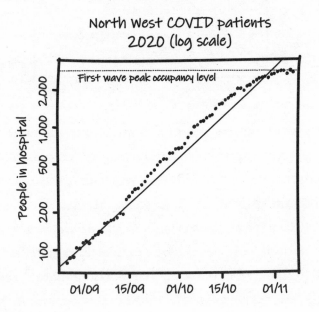

Summary

In this chapter I have introduced a variety of statistics terminology, the language of null hypotheses, p values, statistical significance and confidence intervals. All this can seem a bit daunting, but essentially it's all just a way of making more informed decisions about the world. Rather than going on gut instinct as to whether a new drug seems better, we can use ideas about random variation to calibrate whether the results of the trial are simply the kinds of thing we would expect to see by chance. Again, there is a visual way to think about some statistical questions in the form of regression, which allows us to draw straight lines through data and to think about what they mean: Does this line fit well? Is it possible that the line is not sloping at all? Is it even reasonable to draw a line through this data in the first place?

Suggestions

To take some of these ideas further, I would encourage you to look again for graphs that are published, specifically those with straight lines plotted on them. What is the straight line trying to tell you? Does it look like a good fit to all the data, or is it being unduly influenced by just one or two points? Further, you might like to delve behind the headlines next time you see the results of a clinical trial reported in the news. If a drug seems to work better, can you find the original research paper where the results were reported? You may find a press release from a university which will point you in the right direction. If you do find the paper, there are things you might like to think about. How big was the sample size? What were the p values – is this a 'knock it out of the park' effect where the p value is tiny (say 0.01%) or a very marginal effect (say 4.8%) which could simply be due to random chance and publication bias? How strong an effect does this seem to be?

Chapter 7
Positives and negatives

Context matters

On 4 June 2022, journalist and contrarian Toby Young tweeted: 'Since vaccination in Iceland began, 91% of Covid deaths have been in the vaccinated, but only 90% of the country is vaccinated. Age may play a part here.' In fact, perhaps surprisingly, he was right. Age plays a huge part, but from a mathematical point of view, so does dependence of events. Understanding probabilities in scenarios with dependence is a key skill in making sense of the modern world.

For example, suppose you are shopping for a used car. In assessing whether a particular vehicle might break down, it's pointless thinking about the reliability of the average car on the road. Instead, you consider information such as the brand, model, age and mileage of the vehicle, and move from thinking about all cars to thinking about cars of this specific type. In the same way, your motor insurance company would assess your chances of an accident in the next year by considering information such as your age and record of past claims, not based on the population as a whole.

In fact, understanding dependence of events could literally save your life, by helping you distinguish between information and misinformation around vaccines. I can explain better using UK data, where similar versions of Young's claim can be found. For example, consider a blog post claiming, '80% of recent coronavirus deaths are in vaccinated people, yet only 72% of people are vaccinated. This shows that the vaccines do not help!' The key is that people's decision to get vaccinated is not independent of their risk. We already saw in Chapter 2 that the COVID fatality rate is heavily dependent on age – older people are significantly more likely to die if infected. For that reason, most countries' vaccination programmes were heavily targeted by age, with a rollout prioritised to reach elderly people.

As a result, in the UK around 97% of over-80s received multiple vaccinations. Hence, the right way to think about the blog post is to flip over the claim. The correct framing is that only 3% of the most at-risk group are unvaccinated, and yet 20% of the deaths come from the unvaccinated. Crudely speaking this suggests that unvaccinated people might be six or seven times more at risk. Of course, a proper analysis requires more detail, and involves comparing vaccination and death rates across age groups, looking at outcomes for vaccinated and unvaccinated people separately. However, it should be clear that the original claim is not all that it seems.

In general, we make decisions based on information that we have already. For example, suppose I receive an email saying, 'I have information that shares in Company X are a good buy.' Whether

I act on that depends almost entirely on whether the email was spammed by a total stranger to sucker me into a con trick, or sent as friendly advice by a trusted friend with a track record of successful investments. My prior information regarding the status of the email sender determines how much I trust it.

In fact, whether I even see such a message will depend on a mathematical version of similar considerations. Email spam filters determine whether a message seems trustworthy or can be sent to junk, by scoring characteristics of the email such as IS THE SUBJECT ALL IN CAPITALS, ar3 word5 mi5spelt l1k3 tH15 and so on. These scores are calculated and derived using ideas based on so-called conditional probability, which is the mathematics which drives modern artificial intelligence algorithms and so lies at the heart of the modern world.

Perhaps surprisingly, the same ideas also allow us to understand medical testing better. Since we can assume that no test is perfect, it is impossible to interpret the results of any test without the same kind of prior information – specifically concerning the accuracy of the test and the prevalence of the condition in question. In this chapter, I will explain how to do this.

This is an important and difficult area to navigate. Indeed, some of the answers that we find may seem counter-intuitive. Again, the discussion involves probability, but to understand these issues of vaccines, spam filters and testing we need to go beyond the simple situation of independent experiments such as coin tosses and understand a situation where one event can affect another.

Conditional probability

Consider the following toy example. We have 100 children in a school: 22 of them like both football and chips, 6 like football but not chips, 40 like chips but not football, and 32 children do not like either of these things. We represent all these outcomes in a table as follows:

	Football	Not football	
Chips	22	40	62
Not chips	6	32	38
	28	72	100

For example, by taking sums along the columns and rows of the table we can see that in total 22 + 6 = 28 children like football and 22 + 40 = 62 like chips. This means that the probability that a randomly chosen child likes football is 28/100 = 28%.

However, suppose we are told that a randomly chosen child likes chips. What is the probability that they also like football? We can look at a restricted part of the table – the information given tells us that they are one of the 62 chip lovers, so we need only consider people who lie on that particular row. Since 22 of those children also like football, the probability that a randomly chosen chip fan also likes football is no longer 28% but 22/62, or roughly 35%.

What has happened here? It turns out that the events that a child likes chips and likes football are not independent. Learning one piece of information changes our view of the world and forces us to reassess the odds. Formally speaking, a mathematician would say that the probability of liking football *conditional* on liking chips is 35% (alternatively we might say 'given that the child likes chips'). In fact, we might say that liking chips and football are positively correlated events – knowing that one has happened makes the other more likely.

In terms of the table, learning that the child likes chips allowed us to ignore the numbers in the 'not chips' row, so we effectively moved from a population of 100 children to one of 62. That population lies within the 'chips' row, and the way that we calculated the conditional probability was to divide the entries of the row by its total – so 'chips and football divided by chips'.

It is worth noticing that the answer would not be the same if we asked the question the other way around: given that a child likes football, what is the probability they like chips? The first piece of information tells us that this child is one of the 28 football fans. So, the probability they like chips as well is neither 28% nor 35%, but 22/28 or roughly 79%. In general, there is no reason that these conditional probabilities should be the same.

Bayes' Theorem

In slightly more mathematical language, we might say in this case that the probability of A given B is not the same as the probability of B given A. Here, A is 'this child likes football' and B is 'this child likes chips'. However, a statistical result which is both deceptively simple and extremely powerful tells us that there is a simple relationship between these probabilities. This result is known as

Bayes' Theorem, named after the 18th-century clergyman Reverend Thomas Bayes who discovered it.

The simplest way to see this is to notice that the two fractions above, 22/62 and 22/28, both have the same number on top of them (mathematicians would say they have the same *numerator*). This is not a coincidence, since both these calculations involve the same expression – namely the number of children who like both football and chips. Hence, to move from one fraction to another, we simply need to know the term on the bottom of each (the *denominator*). But these expressions are respectively the total number of chip lovers and football fans, which can be calculated very easily.

In other words, Bayes' Theorem allows us to move from knowing the probability of A given B to deduce the probability of B given A (and vice versa). This has important implications – for example, we could think about A as being 'data we have observed' and B as 'the hypothesis is true'. Often, we know the probability of A given B, since the hypothesis governs how the data is generated, and Bayes lets us deduce the probability of the (more interesting) reverse implication, whether the hypothesis might be true.

For example, returning to our coin problem, we might let A be the event that '5,200 Heads are obtained out of 10,000' and B be the event that 'the coin is fair' (we make some assumptions about what proportion of coins really are fair). As described in Chapter 5, we can calculate the probability of A given B relatively easily, so using Bayes' Theorem we can deduce the more interesting thing, namely the probability of B given A – that the coin is fair given the observed results.

A whole branch of statistics, called Bayesian inference, builds on

Positives and negatives

Bayes' Theorem. These ideas underpin the modern machine learning algorithms (sometimes, perhaps optimistically, referred to as artificial intelligence) which power Siri's speech recognition, allow phones to automatically group photographs by theme and which may one day control your driverless car. Again, we can think in terms of data and hypothesis: for example, event A may be 'a certain sound is received' and event B may be 'person is saying the word fish'. We would need a large amount of training data – that is, a collection of people saying the word fish, to understand the probability of hearing that certain sound when that word is spoken. But then, using Bayes we can flip this over, to deduce the probability that the person said the word fish, given the sound heard by the microphone.

This simplified description of how these algorithms work hides an enormous number of advances in computing power and engineering implementation. However, it gives a rough idea of what is going on behind the scenes when you speak to your phone. I will return to some more implications of Bayes' Theorem in Chapter 8.

Reconstructing probability tables

Looking at the formulas above, we can see that the probability of A given B can be thought of as the probability that A and B both happen, divided by the probability of B (for example 22/62 is 22/100 divided by 62/100). We can rearrange this formula to say that the probability of A and B is the probability of A given B multiplied by the probability of B (for example 22/100 is 22/62 multiplied by 62/100). Often the information is presented to us in this form, and this formula allows us to reconstruct tables like the one above.

For example, suppose we have a university with 1,000 students, three quarters of whom are Science students and one quarter of whom are Arts students. If I tell you that a fifth of Science students missed a lecture today and half the Arts students did, then this is a statement about conditional probability – the chance a student missed a lecture given that they are a Science student, and so on. We can deduce how many students missed a lecture overall, and what proportion of missed lectures were missed by Science students.

Starting with the overall numbers, we can deduce that 750 students take Science subjects and 250 take Arts subjects and can write those figures as the row sums. Then, we need to split the Science students into groups of one fifth and four fifths to deduce the values 150 and 600 on the first row, and the Arts students in half to deduce the values 125 and 125 on the second row. Finally, by summing the columns we can deduce that 275 students missed a lecture overall.

	Missed	Didn't miss	
Science	150	600	750
Arts	125	125	250
	275	725	1,000

Then, either by direct calculation with this table or by using Bayes' Theorem, we can deduce what proportion of missed lectures were from Science students (150/275, which cancels down to 6/11 or roughly 55%). These numerical computations are relatively

Positives and negatives

simple to perform and illustrate how probabilities can be combined together to answer interesting questions. However, we can also understand these calculations, using a visual representation of the probabilities in the following schematic grid:

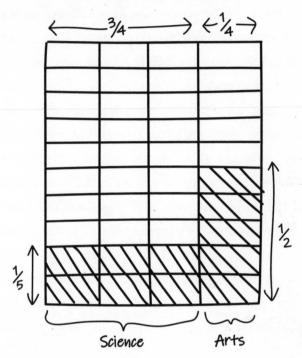

First, we divide the columns of the grid in proportion to the ratios of Science and Arts students. That is, since three quarters of the students study Science, we can divide off three columns to represent the Science students and one to represent the Arts. Second, within each set of columns we can shade in the requisite number of rows to represent missed lectures. That is, a fifth of the Science students missed lectures so we shade in two out of the ten squares in the first three columns, and half of the Arts students did the same, so we shade in five out of ten in the right-hand column.

Now, Bayes' Theorem answered the question, 'What proportion of missed lectures were missed by a Science student?' However, we can use our grid to do the same calculation visually. We consider missed lectures only, so can focus on the shaded squares, so this is really the question, 'What proportion of the shaded squares are in the left-hand columns?' Well, we can count! There are 11 shaded squares, of which 6 are in the left-hand columns, which is where the 6/11 above came from.

This visual method will work in general, even for more complicated collections of events (Science, Arts and Humanities students, say). Since we only care about the proportions of squares that are shaded, there's no need to use the same number of rows and columns that I did. We could have used a grid with 20 rows and 12 columns (say), so long as they were split in the right proportions, and still found the result of 6/11. Indeed, you don't need to use a grid at all – you can simply draw a rectangle and rule off the requisite fractions; all that matters is the ratio of the areas that result. You might like to try this!

Loss functions and symmetry

You will notice that none of this analysis tells us whether a specific randomly chosen missing lecture was missed by a Science student or an Arts student. The best that these kinds of Bayesian methods can do is to give us probabilities, and for us to make a best guess accordingly. Ultimately, in the presence of uncertainty, this is all that we can hope for.

However, it is interesting to think about how we keep score of our guesses. Consider a simple coin-tossing game, where Anne tosses a coin several times, and Bill must predict the outcome of each toss. For each coin toss, if Bill gets the answer right then Anne gives him £1, and if he gets the answer wrong then he must give her £1. It should hopefully be clear that there is no winning strategy that Bill can employ here. As I have mentioned previously, successive coin tosses are uniform and independent, meaning that he can do no more than guess. In the long run, we expect that he will get about half his guesses right, and so will roughly break even.

In more standard betting terminology, we say that Bill stakes £1 on each guess. If he is wrong, he loses his stake. If he is right, his stake is returned to him plus an extra pound. This is commonly referred to as 'betting at evens'. Since Bill expects to break even in the long run, this is referred to as a fair game. (In practice, a bookmaker or casino is unlikely to offer as generous odds as Anne, and Bill would expect to gradually lose his money.) I will describe more about how betting odds can give us insights into probability in Chapter 8.

However, this game is somewhat simplistic, in that there are equal consequences of making a wrong guess in either direction. It is as bad to guess Heads and see Tails as it is to guess Tails and see Heads. We would say that the *loss* associated with those events is symmetric. However, it is not hard to think of events during the pandemic where errors in one direction or another did not have equal consequences, and the loss was far from symmetric.

For example, a healthy patient wrongly diagnosed as infected by a PCR or lateral flow test would have been forced to undergo quarantine and potentially lose salary. On the other hand, an infected patient wrongly diagnosed as healthy would have potentially gone on to infect more people. These consequences are far from equal – and it is not necessarily obvious which is worse when balancing economic and health risks. Indeed, to heavily favour one type of consequence over another may lead to an abundance of caution in one direction or another – not diagnosing anyone as positive or diagnosing everyone as positive.

Another scenario where loss is not symmetric comes in modelling healthcare resources. If we think beds required will be low, but there is great demand, then there will be a potentially catastrophic breakdown in the health service as patients are unable to be treated. On the other hand, if we think that many beds will be required and they are not, then we will invest in surge capacity (such as the UK's emergency Nightingale hospitals, personnel who can staff them and medical supplies) that will prove unnecessary in hindsight.

Of course, these consequences need to be costed, but to me the argument that 'the Nightingale hospitals were never used, therefore the money spent on them was wasted' seems to misunderstand the asymmetric nature of the risks involved. Equally though, we would hope that forecasts would be accurate, and not consistently err in one particular direction.

Medical testing

Few statistical topics around the pandemic created as much speculation and misunderstanding as the issue of false positive test results. For the first few months, coronavirus tests were almost exclusively carried out using PCR testing, which amplifies DNA samples and looks for genetic material associated with the coronavirus. This form of testing is not perfect, but misunderstandings around the issues of test error and prevalence helped contribute to the false 'casedemic' narrative of summer 2020 (the suggestion that the observed rise in cases was not serious, because deaths took longer to start to increase).

Indeed, the problem of inaccurate testing goes well beyond the coronavirus. In most Western countries, people above a certain age will be routinely screened for cancer. None of these tests will be perfect – and may either miss some patients who do have cancer, or wrongly suggest that someone does.

Indeed, there may well be a trade-off between these two effects. Broadly speaking we could imagine a situation where we were extremely cautious and treated all slightly dubious scans as a positive test. Equally, we could imagine a much more blasé policy, where we would only treat a test as positive if we were certain cancer was present. In fact, we could imagine a complete range of possible strategies, based on dialling up or down the sensitivity of the test (the threshold of evidence required to judge it as positive).

Clearly, this level of sensitivity would have an impact on the overall outcomes. So too would the prevalence of the disease, and indeed personal risk factors – for example, we may choose also to screen younger people with a family history of a particular cancer with a known genetic component.

There are no easy answers, and some difficult trade-offs to consider. If we miss the opportunity to make an early diagnosis of cancer, then clearly this has important health consequences through missed treatment opportunities. But equally, if we wrongly tell too many people that they have tested positive, then we will unduly distress them, add to waiting lists, potentially carry out biopsies which themselves carry a very small level of risk, and generally undermine faith in the whole testing system.

Further, the effects of error will vary for different types of diseases: if someone with cancer is wrongly given the all-clear then only their health will be affected, whereas someone with an infectious disease being given the same message may act in a riskier way and infect several other people in due course.

However, now that we have understood conditional probability and loss, we can start to understand properties of medical testing and to calibrate the rates of false positive and false negative errors. Two key properties of any test are its *specificity* and its *sensitivity*, both of which are defined in terms of conditional probability.

The specificity measures how successful the test is for someone who does not have the condition. It gives the proportion of such tests that are negative – so is sometimes referred to as the 'true negative' rate (since such people really should test negative). In the language above, this is the probability of a negative test result, given that the person is healthy. Someone who tests positive when they do not have the condition is referred to as a *false positive*.

The sensitivity measures how successful the test is for someone who does have the condition. It gives the proportion of such tests that are positive – so is sometimes referred to as the 'true positive' rate (since such people really should test positive). In the language above, this is the probability of a positive test result, given that the person is not healthy. Someone who tests negative when they do have the condition is referred to as a false negative.

Positives and negatives

	Infected patient	Not infected
Negative test result	False negative	True negative
Positive test result	True positive	False positive

As they are both measures of success, we would like both specificity and sensitivity to be as large as possible. We can optimise one or the other of them in an essentially stupid way: a test which never even looked at the patient and declared them to be healthy regardless would have 100% specificity (all healthy people would get the right result). However, such a test would have 0% sensitivity (all people with the condition would get the wrong result). We can generate the reverse outcome of 0% specificity and 100% sensitivity by declaring everyone to have the condition. Indeed, tossing a coin to decide the outcome would have 50% specificity and sensitivity.

These toy examples show that we need tests with high values of both specificity and sensitivity. Further, they suggest that there will be some trade-off when tuning a test – it may be possible to achieve higher sensitivity (looking harder for evidence of the condition) at the cost of lower specificity (greater chance of misleading evidence), and vice versa.

It is tempting to think that we should have some measure of overall accuracy, and that a random person be diagnosed accurately with high probability. However, since the prevalence of a disease is typically extremely low, again this is not enough. If 1% of people have the condition, then our stupid 'declare everyone uninfected without even testing them' rule would have 99% accuracy in this sense.

Numbercrunch

Coronavirus testing accuracy

We can now analyse the performance of the PCR tests used to detect the coronavirus. All these calculations could be repeated for other forms of medical testing, given values of sensitivity, specificity and prevalence, but for concreteness I focus on COVID here.

The ONS have estimated that such PCR tests have between 85% and 98% sensitivity – essentially errors arise from an infected person not being swabbed correctly. The specificity is a matter of debate, but ONS suggest it could be as least 99.9%, based on the extremely low proportion of positives observed in the Infection Survey during summer 2020. (If only 0.1% of the survey test positive, even if nobody at all has the condition, then the false positive rate is 0.1% and the sensitivity is 99.9%.)

Overall, we can see that PCR testing is remarkably accurate, in terms of both sensitivity and specificity. However, some issues can still arise due to the low prevalence of the disease, which can be best understood by constructing a table of outcomes, somewhat like the Science and Arts example above – remembering that specificity and sensitivity are expressed as conditional probabilities. We can consider numbers at the more pessimistic end of the ONS range – say sensitivity of 80% and specificity of 99.5%.

Just as in the Science and Arts example, we need one more number, the prevalence of the disease. For concreteness we will assume that this is 1%, and imagine that we test 1,000 randomly chosen people, so we expect about 10 of them to have the disease. The sensitivity value means that 80% of the infected people will test positive, meaning that we expect 8 positive tests and 2 negative tests from that group. Similarly, 990 people will not have the disease, and the specificity means that 99.5% of them, or roughly 985 people, will test negative. (We will round sometimes to get whole numbers.) This leads to 5 false positives. The table of outcomes looks like this:

	Positive test	Negative test	
Infected	8	2	10
Not Infected	5	985	990
	13	987	1,000

As before, we can use Bayes' Theorem or just look at the table to answer the following key question: what is the chance that someone who has tested positive is infected? There are 13 positive tests, 8 of which come from infected people, so 62% of positive tests are accurate – but 38% of positive tests are not. This seems like a surprisingly bad outcome: our gold standard PCR test, with high specificity and sensitivity, nonetheless appears to end up giving positive results that we do not have a high degree of confidence in.

The reason for this outcome is the low prevalence. Since there are very few infected people in the sample, there can be very few true positives. Even if the test had 100% sensitivity, then we would have 10 true positives and 5 false positives, and a third of positive tests would be wrong. This indicates that PCR testing, accurate though it is, has issues when used for mass screening at low prevalence.

However, this is not how PCR testing was generally used in

the pandemic. Its use was mostly restricted to people with symptoms or to contacts of people known to be infected. Indeed, we can use Bayes' Theorem again to see how this affects the calculations. Suppose again that 1% of the population had COVID, and that half of these displayed symptoms such as a fever and cough. Of course, these symptoms are not enough on their own to diagnose COVID, because at any given time perhaps 5% of uninfected people would have these symptoms through other unrelated conditions. However, we can use these numbers to do the same kind of calculation:

	Symptoms	No symptoms	
Infected	5	5	10
Not Infected	50	940	990
	55	945	1,000

The key thing is to look at the column marked 'Symptoms', which tells us that 1 in 11 people (or 9%) showing symptoms have COVID. This doesn't seem very useful to us. However, we wouldn't expect it to be. No doctor would diagnose COVID on the strength of these symptoms alone – although the increase from 1% to 9% for someone showing symptoms may justify the need for them to self-isolate while awaiting a test.

However, by offering PCR tests to those showing symptoms only, we now expect that there are 90 infected people who are tested (this is 9% of 1,000), rather than 10 as before. We can therefore rewrite our original testing table:

Positives and negatives

	Positive test	Negative test	
Infected	72	18	90
Not Infected	4.5	905.5	910
	76.5	923.5	1,000

Now the odds are much better. Of the 76.5 positive tests, 72 of them correspond to people who have COVID. This means that 94% of people who are told by a PCR test that they have the condition will indeed be infected.

In that sense, it appears that the focus on false positives was an overblown concern. By only testing those with a higher chance of being infected rather than by screening at random, the proportion of positive tests that correspond to non-infected individuals is seen to be extremely small.

By making these kinds of calculations, it is possible to design useful tests. For example, from March 2021 the UK encouraged all schoolchildren to take lateral flow tests, which again have a small but non-zero chance of producing a false positive result. Since this represented population-wide screening, with the dangers described above, there was again a possibility that too high a proportion of positive tests might be false positives, particularly at low prevalence. However, data revealed that 82% of positive lateral flow tests passed a confirmatory PCR test in April 2021, already an encouraging fraction. This percentage continued to rise until omicron brought high enough prevalence that the need for a confirmatory PCR test was removed in January 2022.

Indeed, based on the calculations that we have performed, we should perhaps have been more concerned about the false negatives arising from PCR testing: 18 people who are infected would be wrongly given a clean bill of health, which may cause them to act in a riskier manner than they would otherwise do. As mentioned before, the loss function is not symmetric here, and test results which are wrong in either direction can have a quite different outcome in practice.

Of course, my numbers for specificity and sensitivity, prevalence of the disease and percentage of people displaying symptoms are only indicative. They are chosen somewhat in the spirit of Fermi estimation to provide a reasonable guess. Changing the values by a relatively small amount would have little effect on the outcome of the calculations, which would continue to show that mass screening by PCR testing would indeed produce a number of false positives, but that we can have confidence in more targeted testing.

Understanding inequality

The key message of this chapter is the value of understanding conditional probability, and the method by which calculations are performed using it. Many situations arise where outcomes are stated in this way, and it is an invaluable skill to be able to construct tables of outcomes, and to think about what they mean.

For example, being familiar with conditional probability can help you to start to probe the causes of social inequality. Often news reports focus on inequality of outcome, such as one group being disproportionately represented. However, to understand how this has arisen we may wish to focus on inequality of process, which can be fruitfully understood by considering conditional probability.

Consider the following toy scenario. There are two engineering companies, both of whom perform a major recruitment exercise. However, following the process, they are both disappointed to learn that they appointed twice as many men as women, because they hoped to have a workforce that reflected the diversity of the population as a whole.

Only on delving a little deeper into the figures, and thinking about conditional probability, can we start to see that the inequality has arisen differently for the two companies, who would therefore need to address different stages of their hiring procedures.

We can represent the outcomes for each company in a table like this, counting the numbers of men and women who were appointed and rejected. For Company A we see the following table:

	Appointed	Rejected	
Men	20	30	50
Women	10	40	50
	30	70	

By looking down the 'Appointed' column, we can see that indeed 20 men and 10 women were appointed. However, by looking along the rows we can see that the total number of applications received by men and women were equal. In terms of conditional probability, a male applicant had a $20/50 = 40\%$ chance of being appointed, whereas a female applicant only had a 10/50 = 20% chance. This suggests that inequality arose at the selection stage, and that it may be worth examining issues such as possible biases among the hiring panel, or the gendered nature of any assessments that are performed.

In contrast, the table for Company B looks somewhat different:

	Appointed	Rejected	
Men	20	40	60
Women	10	20	30
	30	60	

Looking down the 'Appointed' column, we can see that the outcome of the process was exactly the same. However, looking along the rows and thinking about conditional probability we see a different story. That is, the numbers of applicants were quite different to those for Company A. In fact, the selection process has had a similar effect on each group: in both cases a third of applicants were accepted (20/60 and 10/30 respectively). This suggests that the selection process itself may not be the problem, but rather that inequality has arisen at an earlier stage, with fewer female applicants, possibly because of off-putting gendered messages in the job advert or because of an underlying imbalance in the pool of Engineering graduates.

Of course, this is a simplistic toy example, and in the real world the numbers will never be as clean as this. Further, with small samples of people applying for any role, there will be a degree of natural random variation in any case. However, such an analysis can and should be carried out by larger organisations with more robust data, as a point to reflect on.

This analysis can be extended to cover more stages of the process, for example whether imbalances tend to arise at shortlisting or at interview and so on. This kind of audit is routinely carried out by UK university departments as part of the Athena SWAN gender equality assessment process. While such numbers can never prove anything for sure, understood correctly they can be a good starting point for discussion, and can focus attention on particular parts of an organisation's processes.

Summary

In this chapter, we have seen how many important problems need to be understood in terms of events that depend on each other, using the language of conditional probability. Using Bayes' Theorem and other tricks such as tables and pictures, we can start to make sense of these kinds of settings, including examples such as coronavirus testing and diagnosing inequalities in a simple model of hiring. Using the idea of loss to quantify the idea that different wrong actions can have different consequences, we have seen how to make inference and decisions in a Bayesian context.

Suggestions

To take these ideas further, I encourage you to look for probabilities being quoted in newspaper reports and elsewhere. It is valuable to get into the habit of thinking about whether you are seeing a probability, or a conditional probability – and if so, conditional on what? One thing to particularly watch out for is people (inadvertently or not!) switching between conditional probabilities of the form 'A given B' and 'B given A', without properly checking the calculations using Bayes' Theorem. For example, the difference between 'testing positive given that you have the disease' and 'having the disease given that you test positive' is an important one, and it's well worth watching out for people switching between them or giving a misleading framing. You might like to try reconstructing the kinds of probability tables that I have used or sketching the corresponding diagrams in settings like these.

Chapter 8
Odds and trends

Odds and evens

I just went into a bookmakers and bet £1 at odds of 3 to 1 that the horse Likely Lad will win the 3.30 race at Kempton Park. What does that even mean, and what was I thinking about probability when I did it? In fact, I have chosen to do this because I think that Likely Lad has more than a 25% chance of winning the race. But what is the relationship between all these numbers? Why does 3 to 1 correspond to 25%?

While I have already discussed how mathematicians can use randomness, probability and uncertainty to formulate a view of the world, it turns out that bookmakers and gamblers can also help us understand these things. These are people who apply probability theory for money, meaning they have a strong incentive to get their sums right and are well worth listening to!

In fact, much of the academic discipline of probability theory first emerged out of questions arising from betting. The idea of an expected value, as introduced in Chapter 5, was first introduced by the French mathematician Blaise Pascal to resolve the so-called 'Problem of the Points', which examined what the fair outcome would be if a gambling game were forced to end early. Many more ideas in probability theory, such as the St Petersburg Paradox, gambler's ruin and martingales, arose and gained their names from gambling scenarios.

At a simple level, we will first think about how we move between probabilities and a familiar concept to anyone who has ever gambled, namely betting odds. I will describe the process of gambling in the standard terminology and framework used by UK bookmakers – other conventions exist in other countries. The £1 is referred to as my stake. If Likely Lad does not win the race, I will lose my money. However, if Likely Lad does win, then the bookmaker will return my stake of £1, along with an additional £3 often referred to as my winnings – my stake multiplied by the odds ('3 to 1' means 3 – we can think of it as the fraction 3/1).

If I were more ambitious, I could have placed a larger bet. If Likely Lad does not win, I will again lose my stake, but if it does win then my stake is returned plus the original stake multiplied by the odds. However, since all stakes get multiplied by the same factor, I will stick to £1 stakes for the purposes of explaining what is going on.

Now, bookmakers are not a charity. They will typically make a profit, not just because gamblers aren't always good at assessing probabilities, but also because of the so-called spread on the odds they offer. In general, the bookmakers will offer slightly smaller odds than the analysis below would indicate, meaning that they pay out slightly less money than the probabilities suggest. This means that in some sense the game is not fair, and the odds are slightly tilted in the bookmakers' favour.

We will ignore this and think about a philanthropically minded bookmaker, who resolved to offer fair odds. What would be the fair odds for such a bookmaker to offer for a bet on Heads in a standard coin toss? It turns out that they should offer the odds that we refer to as 'evens', corresponding to '1 to 1' odds in the language above. That is, if I bet £1 on Heads at these odds, if Tails comes up then I will receive nothing, if the coin shows Heads then I will receive my original stake plus £1 winnings, so will receive £2. (This is exactly the setting of Anne and Bill's game in the previous chapter.)

Thinking about the expected value of the bet makes it clear in what sense these are fair odds. The fair coin produces uniformly likely outcomes, and so the amount of money that the bookmaker gives me has expected value $(0 + 2)/2 = £1$, exactly matching the stake that I paid. In other words, my expected overall return from this game is zero. Of course, coin tosses are random and nothing is guaranteed, but if I played this game very many times, the Law of Large Numbers tells me that I should roughly break even.

This is how we determine the fair odds for a particular bet, by thinking about what odds would make the expected overall return be zero. This is a bet which you would be happy to make with a friend, where there is the entertainment of some mild risk, but no sense that the outcome is unfairly slanted towards one person or another.

It is worth thinking about what fair odds for other £1 bets would be, by thinking about our range of biased coins. If I have a coin which shows Heads with probability 1/3, then the fair odds for a bet on Heads will be '2 to 1'. Again, we can see this by thinking that if I see Tails then I will receive nothing, and if I see Heads then I will receive £3 (my £1 stake plus £2 winnings). Hence the expected winnings would be 1/3 x 3 + 2/3 x 0 = £1, exactly matching my stake. In general, if the probability is less than 1/2, then the fair odds will be better than evens – since I am betting on an unlikely outcome.

In contrast, if I have a coin which shows Heads with probability greater than 1/2, then the odds should be lower than evens. For example, if the probability of Heads is 4/5, then the fair odds will be '1/4 to 1', often referred to as '4 to 1 on'. This means my return on seeing Heads will be £1.25 (£1 plus £1/4) and my expected winnings will be 4/5 x 1.25 + 1/5 x 0 = £1 again. For probabilities greater than 1/2, the fair odds will be worse than evens – we often talk about a likely outcome being odds-on in this sense.

Indeed, we can perform this calculation to find the fair odds corresponding to any probability. You might even be able to spot the simple rule that enables us to do this. The key observation is that probabilities are numbers between 0 and 1. Expressed as fractions, this means that the number on the bottom must be bigger than the number on the top. If we know the fraction, we can rewrite it as a pair of numbers: what's on the top, and what's on the bottom minus what's on the top, which I will write as bottom-minus-top for short.

I list the three examples we have seen so far in the following table. In each case you will see that the bookmakers' odds we found can be expressed in the form '(bottom-minus-top over top) to 1'. To do one more example, if the fraction is 3/5, then the top number is 3, bottom-minus-top is 2, and the fair odds are 2/3 to 1, or '3/2 to 1 on'.

Probability (Fraction)	Top	Bottom-minus-top	Bookmakers' odds	Bayes odds
1/2	1	1	'1 to 1'	1
1/3	1	2	'2 to 1'	1/2
1/4	1	3	'3 to 1'	1/3
4/5	4	1	'1/4 to 1' ('4 to 1 on')	4
3/5	3	2	'2/3 to 1' ('3/2 to 1 on')	3/2

Additionally, I list the 3 to 1 odds quoted for Likely Lad. Doing the same calculation, we can see that these odds would be fair in a scenario when this horse's chance of winning was 25% (or 1/4 in fractional terms). These were fair odds with a 25% chance, so if this horse has a higher probability of winning then the game is tilted in my favour (the expected value is positive). Of course, if the true probability is lower than this, then the expected value is negative. And once again, just because the expected value is positive, there is no guarantee that I will win – this simply tells us about the long-run behaviour over repeated races, and we could easily lose all our money in the meantime.

Although mathematicians and gamblers agree about many things, unfortunately we do not agree on the conventions of how to present odds. In the final column of the table, I give what we might think of as mathematicians' odds. These are simply the bookmakers' odds upside down: that is, they are expressed as 'top over bottom-minus-top'. I will describe these as Bayes odds, for reasons that will shortly become clear. To emphasise the difference, I will always give bookmakers' odds in the form '3 to 1', and Bayes odds as a single number such as 1/3.

In many ways, Bayes odds are more natural. In particular, the bigger the probability then the bigger the Bayes odds. However, the two forms are equivalent, we just flip the fraction to go from one to another, so the Bayes odds act as the 'odds-on' version of the standard bookmakers' odds. For example, the final row of the table has bookmakers' odds of '3/2 to 1 on' and Bayes odds of 3/2.

Interestingly, although I have given a rule to go from probabilities to odds, there is a simple rule to go back from odds to probabilities. This rule is that for particular (Bayes) odds, the corresponding probability is 'odds-over-one-plus-odds'. For example, in our Likely Lad case, Bayes odds are 1/3 and this rule gives (1/3)/(1 + 1/3) = (1/3)/(4/3) = 1/4 as we'd expect. Alternatively, for bookmakers' odds, the probability is 'one-over-one-plus-odds', so the bookmakers odds of 3 on Likely Lad convert into the same probability of 1/(1 + 3) = 1/4.

Turing and the Bayesian Enigma

As we have seen, there are simple rules that allows us to convert probabilities into bookmakers' odds and from odds into probabilities. In the language that we used before, the odds and probability are functions of one another (there is a rule that tells us how to go from one to another). Since knowledge of either of these quantities is therefore equivalent to knowing the other, it is reasonable to ask what we gain by this piece of mental gymnastics.

One answer comes with an interesting historical perspective, dating back to the work of Alan Turing, the great British mathematician who helped develop the world's first computers and who formalised the language in which we think about computation itself. As has become better known in recent years, Turing's work at Bletchley Park during the Second World War helped break the German Enigma codes, building on extremely important contributions of Polish cryptographers such as Henryk Zygalski and Marian Rejewski. However, although Turing has achieved a level of popular name recognition which is very unusual among mathematicians, including being the subject of a Hollywood movie (*The Imitation Game*), biographies and a play (*Breaking the Code*) and now appearing on the British £50 note, it is likely that not everyone has thought about what Turing's work involved.

It is natural to imagine that breaking the Enigma codes was something like solving a giant crossword or an enormous jigsaw, a puzzle to be figured out by pure deductive logic. However, Turing's ideas were firmly rooted in probability theory. His first research contribution, which won him a three-year fellowship at King's College Cambridge, was a new proof of the Central Limit Theorem.

However, Turing's greatest contributions to statistics and probability came at Bletchley, and for reasons of national security often only emerged second-hand after his death, through the tireless efforts of his former assistant I.J. (Jack) Good. From reading Good's accounts, it becomes clear that Turing's ideas were founded on Bayes' Theorem, formulated in a more elegant way than we saw in Chapter 7, and involving odds rather than probabilities.

One way to see that this is a natural formulation is to think about the workings of Enigma. Rather than a fixed cipher such as the Caesar Cipher, where A is always encoded as B, B encoded as C and so on, Enigma was designed to generate a vast number of possible ciphers one after another. Each time a key was pressed, an electrical signal was sent through a system of plug-in wires and wheels, to light up a bulb on another keyboard corresponding to a different letter. Essentially each configuration of plug-in wires and each wheel position would generate a different cipher – and each key press moved the wheels around by one step, meaning that successive letters were encrypted in separate ways.

This was a mind-bogglingly complex problem. The machine could be in so many states that there was no chance to simply try them all. Essentially, the only clue came from the fact that letters came in pairs: if pressing key A would light the bulb under letter G, then pressing key G would light the bulb under letter A. As a result of the design of the machine, no letter could ever be matched with itself. However, even given this single piece of information, the position seemed hopeless.

However, Turing realised that the decryption process could be thought of using Bayes' Theorem. Remember from Chapter 7 that Bayes gives us a way to relate events A that look like 'data that we have observed' and B that are 'the hypothesis is true'. Turing realised that he could take the event A, the observed data, to be the output of the Enigma machine, overheard as Morse code by wireless operators. The event B, the hypothesis to be tested, could be something of the form, 'The first plug-in cable connects letters W and F'. Since the wiring of the machine would determine the probability that we saw the output A given settings B, Turing realised that Bayes' Theorem could potentially allow him to deduce the probability of settings B being correct given the observed output A.

Given enough data, we might hope that this conditional probability could point us to the right settings. This may sound simple in theory, but the sheer complexity of the Enigma machine made it bewilderingly difficult to perform this calculation, particularly without a modern programmable computer. However, Turing realised that Bayes' Theorem, usually stated in terms of probabilities, could be reformulated in terms of odds to make this calculation more tractable.

Good's account of Turing's work talks in terms of the evidence in favour of a hypothesis, and he rephrases Bayes' Theorem using a quantity called the Bayes Factor, which tells us how receiving the evidence changes the odds. This formulation of Bayes' Theorem says that, after the evidence is received, the Bayes odds that the hypothesis is true is equal to the original Bayes odds multiplied by this Bayes Factor.

Instead of thinking of all the evidence arriving at once, Turing realised that the calculations could be simplified further by using what was referred to as sequential analysis. That is, each new letter of Enigma cipher was a new piece of evidence, which gave rise to a new Bayes Factor, which could be multiplied together in turn to update the odds.

Alternatively, the calculations may appear even more natural in terms of the logarithm of the odds. Remember from Chapter 3 that

8 x 4 = 32 could also be understood as $2^3 \times 2^2 = 2^5$, with 3, 2 and 5 being the logarithms of 8, 4 and 32, respectively. In general, if we multiply two numbers together, the logarithm of their product will be the sum of the logarithms of the original numbers.

This means that instead of multiplying the odds by the new Bayes Factor, it can be more natural to add the logarithm of the odds to the logarithm of the Bayes Factor. Curiously, Turing and Good came close to thinking of information as a mathematical object in the way that I will describe in Chapter 9. However, unlike Claude Shannon whose work I will describe there, instead of working in units of 'bits', they used what we might think of as an imperial unit of information that they referred to as the 'ban'.

Remember that we hope to find hypotheses with the probability of being true close to 1. This means that we want the number on the top of the fraction to be much bigger than bottom-minus-top, or that the Bayes odds will become extremely large indeed. In other words, we hope to collect successive Bayes Factors that are big enough to make this happen.

Specifically, Bayes' Theorem was used to compare the weight of evidence in favour of a variety of hypotheses. Remember that Enigma machines step through a sequence of codes, meaning that 'successive letters were encrypted in separate ways'. However, when two Enigma machines were used with their initial wheel positions in a similar state, the sequence of codes would be the same for each machine, just offset in time one against another (one machine may be seven steps above, or 28 steps behind, for example).

It was possible to test this by finding the weight of evidence for the hypothesis that the machines were one step apart, two steps apart and so on. This evidence was formed by coincidences in the output of the machine, which arise because English or German messages are somewhat predictable, since for example the letter E is more common than the letter Q. (We will see more about predictability in language, as measured by my academic hero Claude Shannon, in Chapter 9.) As a result, if we have correctly guessed the offset then the machines will be in the same state, and the chance is somewhat larger than 1/26 that they will be encoding the same letters and hence will produce the same output.

By looking for these coincidences, counting their occurrence using specially produced sheets of paper, and calculating the logarithm of the Bayes odds, the Bletchley team could suggest possible offsets between pairs of Enigma machines, which could be tested further using the automated 'bombe' devices. In this way, the Allies could gradually deduce the settings of the Enigma machines, read the messages that were being sent, and act on the intelligence received, helping to ensure the success of the D-Day landings.

Turing's thinking foreshadowed much of modern statistics, which as I have described is dominated by Bayesian inference, and of course by the enormous computing power which has grown out of his original ideas to build physical devices to solve abstract mathematical problems.

Bayes Factors and false positives

To illustrate further how the Bayes Factor method can be useful, we will return to medical tests, as discussed in Chapter 7. We will see how the calculations become much more natural when stated in terms of odds rather than probabilities.

Numbercrunch

The piece of evidence that we will consider is the fact that we have tested positive for a disease. Before the evidence arrived, there was a certain probability that we had the disease, which we can convert into odds. We refer to these as background odds, acknowledging the fact that they tell us the situation before we learn anything. Our background odds can simply be based on the overall population prevalence of the disease, but ideally would aim to integrate personal factors such as a genetic predisposition towards the condition, lifestyle issues such as obesity and smoking, and short-term markers such as showing symptoms or having been a recent contact of an infected individual.

Remember that, in Turing's language, the Bayes odds that we are infected given that we tested positive are the background Bayes odds that we are infected, multiplied by the Bayes Factor. For this disease testing example, it turns out that the Bayes Factor is the probability of testing positive given that we have the disease, divided by the probability of testing positive given that we don't.

To understand why, we can return to our example of Science and Arts students missing lectures from Chapter 7. You might recall that at our fictitious university, three quarters of students studied Science and the rest Arts, and that a fifth of Science students and half of the Arts students missed their lecture, which we represented like the chart on the following page:

Turing's statement tells us that the Bayes odds that a student studies Science given that they missed a lecture equal the background Bayes odds that they are a scientist multiplied by the Bayes Factor. For this university example, it turns out that the Bayes Factor is the probability of missing a lecture given that they

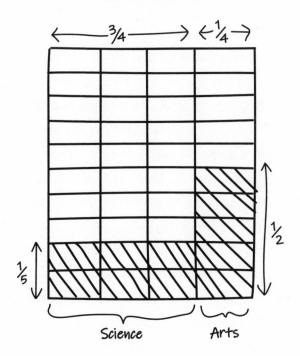

study Science, divided by the probability of missing a lecture given that they study Arts.

This might seem daunting, but we can check how this formula works by plugging in the numbers that we already have. The Bayes odds that a student studies Science is 3 (the probability is 3/4, so our 'top over bottom-minus-top' rule gives '3 over 4-minus-3', or '3 over 1'). The Bayes Factor is 1/5 (probability of missing a lecture given that they study Science) divided by 1/2 (probability of missing a lecture given that they study arts), or 2/5. Hence Turing's rule tells us that the Bayes odds we are looking for are 3 times 2/5, or 6/5. This perfectly matches the calculation of the previous chapter, where we found that the probability of being a Science student given that you missed a lecture was 6/11, which again by our 'top over bottom-minus-top' rule gives Bayes odds of 6/5 as we'd hope.

Numbercrunch

This example also helps us understand where this Bayes Factor method comes from, and why it works. That is, the odds of 6/5 simply represent the ratio of the number of shaded squares on the left of the line to the number of shaded squares on the right. The 6 arises as the product of 3 columns and 2 rows, the 5 is the product of 1 column and 5 rows.

So, we can break up our fraction as 6/5 = (3 x 2)/(1 x 5), or if we prefer as (3/1)x(2/5). The first term gives the background odds, the second term gives the Bayes Factor.

Having (hopefully!) understood this argument, we can plug in the numbers from Chapter 6, and verify the calculation performed there for random testing for coronavirus. Remember that we assumed that 1% of the population were infected, 80% of infected people would test positive, and 0.5% of the non-infected people will test positive.

If we test people at random, then the 1% of infected people correspond to a fraction of 1/100, so top is 1, bottom-minus-top is 99, giving background Bayes odds of 1/99. Remember that the Bayes Factor is the probability of testing positive given that we have the disease, divided by the probability of testing positive given that we don't. So, in this case we have 80/0.5, or a Bayes Factor of 160.

So, multiplying these numbers together, we get that the Bayes odds we are infected given that we tested positive are 1/99 x 160, or roughly 8/5 (rounding for simplicity!). This value matches the odds we found in Chapter 7 by a more involved method. There we found that 8 out of 13 people testing positive would be infected, and our usual 'top divided by bottom-minus-top' rule converts this 8/13 fraction into Bayes odds of 8/5, meaning that the two methods agree.

Odds and trends

This may all seem daunting. Again, we can see the answer more clearly by drawing a picture. I have exaggerated the scale slightly to make the regions visible, but the following figure represents the probabilities in question. The leftmost 1% of the square is ruled off to represent the infected people, 80% of that left-hand rectangle is shaded to represent those infected people who test positive, 0.5% of the right-hand rectangle is shaded to represent uninfected people who test positive.

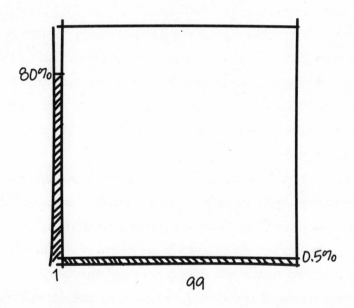

Hence, the odds we care about represent the ratio of the areas of the two shaded rectangles, one tall and thin, the other short and fat. The first rectangle has area 1 x 80 = 80, the second has area 99 x 0.5 = 49.5, so the ratio of their areas is 80/49.5, again our 'roughly 8/5'. If you prefer, then once again we can write the fraction differently. Rather than thinking about it as 80/49.5, we can write it as (1 x 80)/(99 x 0.5) = (1/99)x(80/0.5) = (1/99) x 160. Once again, the first term gives the background Bayes odds, the second term gives the Bayes Factor.

This is a slightly long-winded way to quantify the fact that PCR testing for coronavirus worked well. If you prefer to think of it in bookmaking terms, learning that a PCR test is positive is like insider information that turns a 99-to-1 outsider into an odds-on favourite. As at Bletchley Park, we were hoping for large Bayes Factors that would multiply up relatively small background Bayes odds to provide much larger ones, and this is exactly what PCR testing gave us. Indeed, looking at the form of the Bayes Factor, we can see how this large value of 160 is arrived at: a high proportion (80%) of people who should test positive do, and this is divided by an extremely low proportion (0.5%) of people who shouldn't test positive but do.

In an extreme example, if the false positive rate was 0%, then we would divide by zero to provide a Bayes Factor of infinity. This is the best possible result and fits with the intuitive understanding that everyone who tests positive in these circumstances really would have the disease. In an extreme example in the other direction, we could imagine a test which had a Bayes Factor of less than 1. In that case, a positive test would reduce the odds that someone has the disease, meaning that the test is providing us with misinformation! Overall, this Bayes Factor gives a useful way to measure test quality: the larger the Bayes Factor, the better the test.

It may feel like we have not gained much so far. We've simply converted one set of calculations into another and checked that we got the same answer as in Chapter 7. However, the real benefit comes when we repeat the calculation. The key is to notice that the Bayes Factor only depends on the test itself, meaning that it is a universal quantity. If I perform the same test in Warrington and Wellington, the Bayes Factor will be the same. The only thing that changes will be the prevalence, expressed through the background odds.

For example, in Chapter 7 I repeated the calculation of false positives based on testing the group of people with symptoms, rather than testing at random. In that case, where the prevalence in that group was 9%, the background Bayes odds are 9/91 (9% is 9/100, so top is 9, bottom-minus-top is 91). We simply multiply this by the same Bayes Factor of 160 to obtain Bayes odds of (9/91) x 160, or about 16. This matches the calculation of Chapter 7 which gave 72 true positives and 4.5 false positives, corresponding to the same Bayes odds.

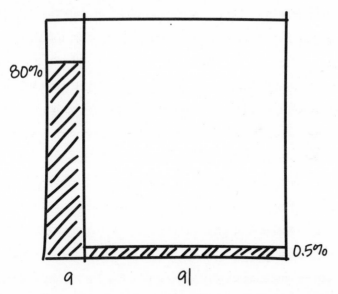

In terms of our picture, we simply slide the dividing line further to the right to produce a wider shaded rectangle on the left-hand side, with a correspondingly larger area, and correspondingly larger odds. The proportion of each rectangle which is shaded does not change, remaining at 80% and 0.5% respectively, because the conditional probabilities are an intrinsic property of the tests themselves.

This picture shows why the Bayes odds are higher when we only test people with symptoms, because they represent the ratio of the area of the tall rectangle to the flat one. As the left-hand rectangle is now nine times wider, its area has grown by the same factor. The right-hand rectangle has roughly the same area as before. So roughly speaking, the Bayes odds increased by a factor of about nine as a result of only testing people with symptoms.

This is an extremely attractive benefit of the Turing-style odds-based formulation of the false positives problem, which allows us to separate out the test itself (expressed through the Bayes Factor) from the background odds of the population that we test. In fact, this Bayesian formulation gives a beautiful result: double the prevalence, double the odds. Halve the prevalence, halve the odds.

Indeed, this allowed me to tell that there was something wrong with the UK's COVID testing system in the autumn of 2021. At a time when prevalence was rising, the odds (as measured by the proportion of lateral flow tests passing PCR confirmation) actually fell, which was a result that made no sense. This proved to be a sign of problems in one particular lab and monitoring these kinds of metrics can help detect any such problems in the future.

In fact, while I have described this result in terms of population level testing, the same argument works for individuals undergoing medical testing. Imagine testing individuals for a certain cancer, which has several known risk factors, such as age, smoking and alcohol. We would like to combine these risk factors and the result of a test into an overall assessment of the probability that they have cancer.

Odds and trends

Clearly, Bayes' Theorem is the right way to do this, but equally we cannot expect every doctor to have a working knowledge of statistics sufficient to perform this calculation, nor the time to look up test success rates and do the requisite mathematics in the middle of their working day. However, the odds formulation of Bayes' Theorem allows them a way to find this risk easily.

Remember that the odds of having the disease following a positive test are the background odds for that person multiplied by the Bayes Factor, and we can rewrite this to say that the logarithm of the odds is the logarithm of the background odds plus the logarithm of the Bayes Factor.

The key is to remember that the Bayes Factor is the same for every test of this kind, so it can be declared as a standard value in advance, perhaps by the medical regulator. That means that every time the doctor performs this test, they simply need to perform a pre-test assessment of the patient's risk factors to find the background odds, and convert this into the odds of the disease following a positive test.

We can represent this on a graph as follows:

We simply need one graph for each kind of test we perform, with the requisite logarithm of the Bayes Factor marked on it. Then, the doctor can simply find the patient's background odds, find that point on the x-axis, look up to the sloping line and read off the odds from the y-axis. In fact, it may even be simpler than this in practice: we may simply decide some appropriate threshold for the odds – such as 'odds above 0.2 mean that further tests are required' – which would convert into a threshold for the background odds. In other words, the doctor would not even need to

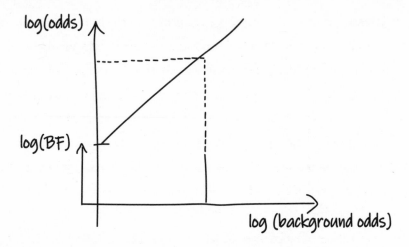

look at the graph: anyone testing positive with background odds above a certain level should be invited for further tests.

Fractions and spread

There are other contexts where thinking about odds instead of fractions can help us understand how natural processes behave. In many contexts, we will see what are often referred to as S-shaped, logistic or *sigmoid* (from the Greek for the letter S) curves. I give an example of one of these curves below, simply generated as a toy model on a computer:

The key is to think about this as a graph of percentage market share. It represents a new product moving from almost no share, via a very steep growth phase to achieve almost complete saturation of the market. Certain phases of this curve can be deceptive: early on it looks as if it is growing exponentially (and indeed this is roughly true). However, clearly exponential growth cannot be sustained forever, not least because there is a hard upper limit of 100% market share.

Odds and trends

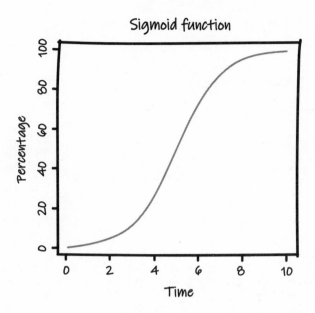

However, this curve does not simply grow exponentially up to the 100% level and stop. Instead, it gradually flattens out. We can think about the reasons for this in terms of development of a new business. Early on, it is easy for a product to increase its market share; given sufficient buzz, there are plenty of new consumers to take it up. However, as time goes on this becomes harder to sustain. Once a product achieves say 50% market share, it can only increase its sales by converting the remaining 50%, who may well be more resistant due to age, income or stubbornness.

For this reason, often these sigmoid curves do not fully saturate the market – for example, we can look at graphs of percentages of mobile phone users who own a smartphone, or percentages of web users who browse using Google Chrome and see a flattening before they achieve 100% market share. Perhaps a certain proportion of web users simply buy a computer that comes with Microsoft or

Apple's own browser and do not feel the need (or lack the techno-logical confidence) to install a new one.

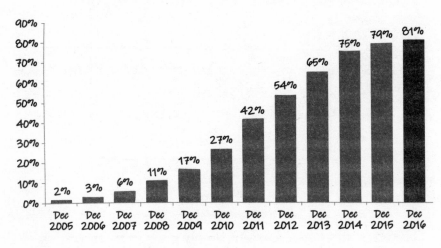

Smartphone Penetration of Mobile Phone Market

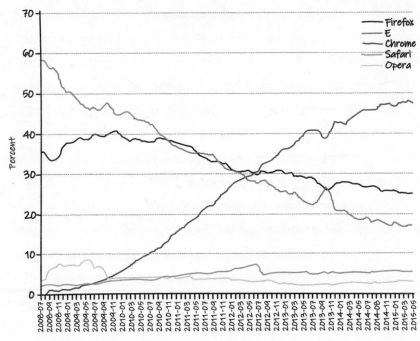

Odds and trends

This kind of sigmoid growth is the kind of behaviour exhibited by COVID variants such as alpha, delta and omicron as they took over in different countries. These variants started with an extremely low percentage of cases, but rapidly outgrew the existing strains to represent almost 100% of COVID tests sampled.

We can understand how this kind of sigmoid curve can arise to predict and model its future behaviour, which is simpler and more predictable than it might look. If we return to the toy model above, instead of thinking about it as a percentage or a proportion, we should think about the Bayes odds. As before, we can calculate 'top over bottom-minus-top' and plot that instead. As market share increases, the Bayes odds will increase, but what kind of curve should we see?

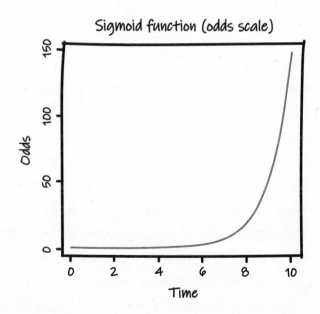

In fact, plotting the odds corresponding to the toy model gives a recognisable shape, which starts flat but gets steeper, suggesting exponential behaviour. As in Chapter 3, we can test whether that is the case by plotting the odds on a log scale. In this case we see a straight line, showing that the odds do indeed grow exponentially.

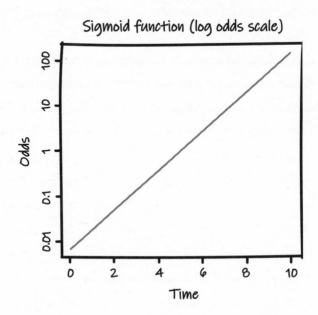

This makes sense when thinking about the spread of a new COVID variant, for example. If we have two strains, types A and B, the proportion of the faster-growing strain will be the number of type A, divided by the number of type A plus the number of type B. This isn't a very nice expression. However, we can convert it into odds in our usual 'top over bottom-minus-top' way to find that the odds are simply the number of type A divided by the number of type B. Hence if both strains are growing or shrinking at an exponential rate, then the odds will behave exponentially, because the numbers on the top and bottom of the odds are each multiplied by a (different) constant factor in each period.

In Chapter 3 we saw that plotting data using a log scale y-axis allowed us to estimate the rate of exponential growth. Similarly, plotting the odds using a log scale y-axis allows us to estimate the difference between the growth rates of the two strains. In other words, this kind of plot (sometimes referred to as a logistic plot) can determine the competitive advantage of a new strain.

It was precisely these kinds of calculations which enabled scientists to realise that variants such as alpha, delta and omicron must be spreading faster, since a straight line appeared when the logarithm of the odds was represented on a logistic plot. This was later confirmed when scientists produced the same kinds of plots for countries around the world, showing that these strains' competitive advantages seemed consistent in many locations (the slope of the logarithm of the odds was roughly the same).

Indeed, plotting proportions or fractions in this way will generally turn a slightly hard-to-interpret sigmoid curve of data into something approximating a straight line. We could carry out the same kind of transformation for the browser dataset, first converting into odds and then plotting the logarithm of the odds, to judge how fast Google Chrome is growing and to extrapolate how fast it might grow in the future.

It's interesting to see two themes of this book coming together in this way. First in Chapter 3 I argued that plotting the logarithm of data can be useful and, in this chapter, I have argued that odds are often a natural way to think about fractions. Putting these together it appears that the logarithm of the odds is an interesting quantity in its own right, and that tracking it can give insights into medical tests and the spread of COVID variants and new consumer products.

Numbercrunch

Summary

In this chapter, we have seen how thinking about probability in terms of gamblers' odds can give us new insights into the likelihood of events. For example, we have seen how Alan Turing's work at Bletchley Park and the question of false positives in medical testing can both be usefully understood in this setting. Further, the trick of converting odds into probabilities, combined with the idea of exponential growth, allows us to predict the spread of variants and to straighten the line on these kinds of plots.

Suggestions

You might like to try playing with some of these ideas yourself. You can look for odds that are quoted on bookmakers' websites and elsewhere and try converting them into probabilities using the 'odds-over-one-plus-odds' rule that I described. Do the resulting probabilities seem plausible to you? If not, does it suggest that a bet in one direction or another could be profitable? (You don't have to go as far as placing a bet of course, you can simply treat this as an exercise to see how well calibrated your judgement is.) You might also like to try carrying out some of the same calculations with odds for medical testing by drawing a picture or calculating a table. For example, if the true and false positive rates of lateral flow tests were 50% and 0.03% respectively, what would that tell you about the odds of a positive test being correct, at 0.1%, 5% and 10% prevalence?

PART 3:
INFORMATION

Chapter 9
Information is power

Claude Shannon

On 23 February 2020, the results of an interesting opinion poll were published by CBS News in the United States. Rather than asking voters whom they would vote for in the autumn presidential election, the pollsters asked them to predict who would win. Although this was a simple assessment of probability, it was hugely divided among party lines: 90 per cent of Republicans thought President Trump 'definitely or probably' would be re-elected, but only a third of Democrats did. What can explain such a disparity?

This is an example of a filter bubble. People assess the likely winner of an election by thinking whether a candidate seems popular and do that at least in part by thinking about people they know. For example, if many of your Facebook friends are vociferous Trump supporters, you might think that was a representative sample of the population. The polarisation of US politics, as well as stratification of society by age and social class, means that Republicans are likely to be friends with Republicans and Democrats to be friends with Democrats. Hence each group will receive a deceptive view of the world and believe that their own views are more popular than they are.

This effect can be dangerous when it contributes to a belief that an election result must be wrong. It is therefore important to understand how the problem of people receiving misleading information about the state of the world can arise. I have argued so far that mathematics is a powerful tool for plotting data in the right way, tracking trends and understanding estimation errors. It is perhaps more surprising that maths has something to say about reconciling apparently contradictory facts, sifting information and making sense of a barrage of data.

But what do we mean by information in this sense, and how can we measure and quantify it? It all began with the work of a single genius, Claude Shannon, who published a paper in 1948 which established a whole new field of study, known as Information Theory, out of the blue. In fact, when I refer to Shannon as a mathematician, some people would argue that he was really an engineer. This is perhaps true, but his ground-breaking work was entitled 'A Mathematical Theory of Communication', and so I will happily claim him as a mathematician.

Shannon is my personal academic hero, and part of his appeal to me is that he never took himself too seriously. He liked to tinker with physical devices: building juggling robots and toy mice that could explore mazes – all this at a time when the few computers in the world were the size of a room. It was Shannon's grounding in building things that helped him make the first breakthrough of his career, when as a 21-year-old student he produced what is generally agreed to be the greatest Master's thesis in history. This work of 1937 established a fact that we now take for granted: that any computational problem that can be represented in terms of 0s and 1s (sometimes known as Boolean algebra, after the Irish mathematician George Boole) could be performed by electrical circuits.

During the war, Shannon worked on classified government problems in cryptography (understanding secret codes), including proving that the so-called one-time pad was an unbreakable code when implemented correctly. One way to think of the one-time pad is in terms of flipping zeroes and ones according to the result of a sequence of independent fair coin tosses.

For example, suppose that the British embassy in Paris want to send the message 01101011, perhaps meaning 'all well here', back to London. The ambassador and her London base can do this securely if they both have a copy of the same one-time pad. This is simply a pre-agreed sequence of zeroes and ones, for example created by repeatedly tossing a fair coin and writing a 1 for each Head and a 0 for each Tail.

A 1 in the one-time pad means 'flip the bit in this position', a 0 means 'leave it alone'. So, if the one-time pad starts 10101100..., it means 'flip the 1st, 3rd, 5th and 6th bits of the message'. Since the message we want to send is 01101011, the signal transmitted will be **11000**1**11**, marking the flipped bits in bold for emphasis. When the London base receives this message, they simply take their own copy of the one-time pad and flip the relevant bits (1st, 3rd, 5th and 6th) back to recover the original message.

However, from the point of view of an eavesdropper, the position is hopeless. If they intercept the transmitted signal 11000111, they have no way of telling which bits have been flipped. The transmitted signal is just as compatible with the message 01010110 ('French secrets successfully stolen') or 11001101 ('send more teabags'), because there are perfectly valid and equally likely one-time pads corresponding to those messages. Assuming that the one-time pad really was generated by independent fair coin flips, it really has been used just once and that nobody has managed to break into the embassy and photocopy it, it is mathematically impossible for an eavesdropper to tell which message was sent.

The key insight is that the process that generated the one-time pad is completely unpredictable, and so can perfectly hide the transmitted message. This realisation was to prove a major part of Shannon's development of Information Theory, because from there it is natural to think about the effect of a one-time pad formed by more predictable coin flips – if we only flip 1 bit in 10, can we recover the message? It turns out that we can do this if we exploit the structure of the messages that are being encoded, just as we can often spot the answer to a crossword puzzle from having some letters in the grid already. For example, if the message 01101011 ('all well here') is often sent, then if we see something close to this because of the action of the one-time pad, then we can be confident this was the message encoded.

During the war, Shannon met another great genius who worked at the interface of mathematics and physical devices, namely Alan Turing, whose work I have already discussed. There is a great degree of commonality between Shannon and Turing's approaches to problems, and they both abstracted written language into mathematical models to deal with problems arising from wartime code-breaking. However, although they met during Turing's 1943 visit to the United States, they never published together, and perhaps due to the terms of secrecy they both worked under may never even have realised the commonality between their work.

Shannon spent a key part of his career at Bell Labs, a research facility which operated somewhere between the immediate commercial pressures of industry and the research freedom of academia. This somewhat freewheeling environment played a key role in the development of key 20th-century technologies such as the transistor and lasers, and work at Bell Labs led to the award of nine Nobel Prizes. Bell Labs was a perfect match for Shannon's blend of hands-on tinkering and fundamental research.

Following the war, Shannon published his masterpiece, spread across two issues of the Bell Systems Technical Journal in 1948. In many ways, this paper helped create the modern world by establishing the way in which information should be quantified. Indeed, Shannon coined the term *bit* (short for Binary digIT) for a quantity that could be 0 or 1.

Shannon realised that these bits were the fundamental units of storage: any amount of information could be represented as a sequence of 0s and 1s. For example, given 4 bits, there are 16 possible messages we could represent by them (there are 2 possibilities for the value taken by the first bit, 2 for the second, 2 for the third, 2 for the fourth, and 2 x 2 x 2 x 2 makes 16). We just need a table that tells us which message is associated with each sequence.

This idea of storing data in 0s and 1s is fundamental today, even if we now think in somewhat larger units. Text is often stored in a byte of data (8 bits), and we now often talk in terms of megabytes (8 million, or 8,000,000 bits), gigabytes (8 billion, or 8,000,000,000 bits) or even terabytes (eight trillion, or 8,000,000,000,000 bits). Every time you buy a hard drive measured in terabytes or a phone contract measured in gigabytes of data, you are using terminology that grew out of Shannon's work. The engineering specifics of how billions of 0s and 1s can be stored on a tiny device might boggle even Shannon. However, nearly 75 years later, his insights still help us understand the fundamental limits of storage and communications systems.

Shannon and entropy

We can understand some of Shannon's contributions by returning to the object that we used to introduce probability: the coin. In fact, think of tossing two coins, one fair one and one which is very biased and produces Heads 89% of the time.

It should be clear that the outcome of tossing the fair coin is essentially unpredictable, being just as likely to be Heads or Tails, and with the previous outcomes having no effect on the next. This means that no prediction strategy can ever hope to be right more than half the time, which is partly why tossing a coin can be used to settle intractable disputes. In contrast, it is relatively easy to predict the outcome of tossing the biased coin – most of the time the answer will be Heads. By always guessing Heads we can expect to be right about nine times out of ten, a much more impressive level of prediction accuracy than for the fair coin.

What Shannon realised was that not only was the biased coin easier to predict than the fair one, the outcome of a sequence of its tosses could be summarised in a more efficient way. That is, if we had to report the result of 128 tosses of the fair coin, we couldn't do better than simply to list the whole set of outcomes HTTHTHTT . . . HH (or whatever it was). If we used Shannon's language, we might represent this in bits, with Heads represented as 1s and Tails represented as 0s. Shannon realised that to summarise the 128 tosses of the fair coin would take 128 bits, or 1 bit per toss.

In contrast, the fact that the biased coin is more predictable gives us ways to save space when describing the sequence. For example, we don't need to list the whole sequence of Heads and Tails, but only need to convey which tosses came up as Tails. So, we could give a sequence of numbers: 11, 18, 32, . . ., 97 – these being the number of the tosses that gave us Tails.

We'd expect there to be about 14 such tosses, and we can represent each one using seven bits (just as before, there are 2 x 2 x 2 x 2 x 2 x 2 x 2 or 128 possible sequences of seven bits, so we can associate each one with a different possible coin toss that could have been Heads). So, on average we'd expect to only need 98 bits using this scheme, or 0.77 bits per toss. Hence, the outcomes of tossing the biased coin can be summarised more succinctly than the fair one.

In fact, Shannon realised that the predictability of the outcomes and the ease of summarising them were equivalent properties. He introduced a new metric, called *entropy*, which measures how random a quantity is. Furthermore, the natural unit in which to measure entropy turned out to be his bit. The question of 'how predictable is a random quantity?' amounted to 'how efficiently can we summarise the result?', and the answer was quantified by 'the entropy of the outcome'.

Shannon gave a formula for the entropy in terms of the probability, and we can calculate the value of this formula for the coins above. It turns out that the fair coin has entropy 1 bit, and the 1 bit/toss is the most efficient representation that we can achieve. In contrast, the biased coin has entropy 0.5 bits, and there are more efficient methods of describing the outcomes than the 0.77 bits/toss described above.

This idea of being able to efficiently represent random objects as a sequence of 0s and 1s is referred to as *data compression*. Essentially, we look for redundancy or predictability in outcomes and remove it to provide a representation requiring fewer bits. This is exactly what your phone camera does when it takes an image which should take hundreds of millions of bits to store and creates a jpg file that uses a few per cent of that. The key is that images are predictable – one blue pixel may be part of a large region of blue sky, and so neighbouring pixels are somewhat predictable, like the biased coin.

However, Shannon showed that, just as we cannot compress all the air in a room down to nothing because the molecules take up a certain amount of space, there is a fundamental limit, given by the entropy, beyond which data cannot be compressed. For example, Shannon showed that no representation of the biased coin could average fewer than 0.5 bits/toss.

Come on feel the noise

Even quantifying the randomness behind data compression would have earned Shannon a place in the mathematical hall of fame. However, his 1948 paper went on to make another huge contribution, by understanding the effect of noise on messages. We might naively imagine that our communication channels are perfect. If we send a letter, we expect it to arrive intact and unaltered at the recipient's house, ready for them to read.

However, electronic communications are not so simple. We all carry round a mobile phone, which we expect can transmit all our conversations perfectly. However, the phone has a relatively small battery and needs to communicate by radio with the nearest mast. Often such communication takes place in a crowded city block, where radio signals will bounce off buildings and where hundreds of phone users may be trying to use the same mast. Thought of like this, it may seem like a miracle that any mobile phone call is ever successfully received. Certainly, it seems natural to assume that the signal transmitted by the phone will not be received perfectly by the mast. We say that the communication channel from the phone to the mast is a noisy one and model the effect of noise by imagining that random errors are introduced in transmission.

Shannon realised that this need not be a problem. Just as he proved the mathematical limits of how efficiently data could be compressed, he showed there were fundamental limits to how successfully information could be transmitted in the presence of noise. He introduced the idea of the *capacity* of a noisy channel. We can think of this as 'how much information can get through it', somewhat like the way that the width of a pipe constrains how much water can pass along it.

Shannon showed that communication was always possible with a low probability of making a mistake, so long as we don't exceed this capacity. The idea was something like our biased one-time pad, where the fact that only a few bits were flipped could allow us to reconstruct the original message. What is more, Shannon was able to quantify this capacity in terms of the same entropy that allowed him to make sense of the data compression problem.

Just as data compression works by stripping out redundancy, Shannon realised that the key to communicating through a noisy channel was to add in redundancy to protect the information in the message. The capacity essentially tells us how much redundancy we need to add, in the form of so-called check bits that allow messages to be corrected by the receiver.

While Shannon's work showed the theoretical limits of what was possible, it took 50 years for practical schemes to be designed that could achieve the performance he predicted. Designing such error-correcting codes that work well in real-life scenarios remains a highly active topic of research today.

Multiple sources of information

We have already seen that Shannon's quantification of uncertainty through entropy plays a vital role in understanding many communications problems. It captures how much we can compress a string of independent coin tosses and how transmitters can send messages that can still be understood by receivers despite the presence of noise. However, entropy has several further properties that are relevant to our understanding of the world.

One key thing is that, as well as capturing the idea of the compressibility of a message, the entropy tells us how surprised we are when we receive it, which in turn corresponds to how much we learn by reading it. So, for example, we learn more that we didn't know already about the outcome by hearing the results of the fair tosses

HTTHTHTT ... HH than we would do by hearing a sequence of the unfair tosses HHHHHHHHHTHHHH . . . Essentially, we already knew that most of the unfair tosses were likely to be Heads, so most of what we hear has been priced into our understanding of the world already.

Another way to think about this is that we learn more from rare events than we do from common ones. The fisherman who in 1938 captured a living coelacanth, a species of fish thought to be long extinct and known only through fossils, changed our understanding of the world far more than the one who pulled out a herring from the North Sea yesterday. Although, by definition, rare events do not happen as much as common ones, when they do happen, we learn much more from them.

Although I have described how much we learn from the sequence of coin tosses, this only represents one simple scenario. That is, as discussed before in Chapter 5, successive coin tosses are independent (one result has no effect on the next one). This independence is not generally the case for successive pieces of information that we receive, which usually come with some degree of correlation between them.

For example, we expect the number of coronavirus deaths in the UK on two successive days to be relatively close to each other, because they are both somewhat determined by the current underlying number of infections. In contrast, numbers of deaths several months apart will be much closer to independent, because there are many possibilities for how the epidemic could have developed in the intervening time.

Shannon showed that, all things being equal, we learn the most from two successive pieces of information that are truly independent. Suppose one piece of data contains 1 bit of information and another piece of data also contains 1 bit of information. If they are independent, then we have 2 bits of information altogether. But if they are not independent (like data from the same place on consecutive days), we get fewer than 2 bits all together – the whole is less than the sum of its parts – because some of the information has 'overlapped'. So, the best information is independent information: because independent bits add up.

Another way to think about this is when information is not independent, having learned the first data point, we learn less added information from the second data point than we would from an entirely independent piece of data. Some of the surprise has been taken away.[8]

Entropy and pooled testing

Shannon's ideas also help in detecting disease through the idea of pooled testing, sometimes referred to as group testing. Since medical tests can be scarce and expensive, we want to use each one as efficiently as possible. If the prevalence of a disease was 1%, then most tests performed would come back negative. As with the biased coin above, this means that the entropy of the tests is extremely low, and that we learn little information from each one, in the sense that Shannon described.

---

8 We can deduce the formula for information from knowing that the information from independent events adds up. We know that the probabilities of independent events multiply, and we know that the logarithm of a product of numbers is the sum of the logarithms. Putting this together, this suggests that the information we gain from knowing an event has occurred should be the logarithm of the probability of the event. This idea actually predates Shannon and goes back to the work of the American engineer Ralph Hartley in 1928. Putting this together with an idea from Chapter 5, Shannon's entropy is the expected value of the information we gain.

Information is power

It may seem that we can do little about this, since we cannot alter the percentage of people who are infected (nor indeed would we want to see more infections taking place). However, we can use a clever idea of an economist called Robert Dorfman, another researcher motivated by problems with military applications during the Second World War. Curiously, although Dorfman's paper appeared in 1943 and hence pre-dated Shannon's 1948 masterwork, it is easier to motivate his ideas using Shannon's language.

Dorfman was involved in the syphilis screening of men enlisted to the US military. While a test existed for syphilis, it was expensive to administer, and the condition was rare. Dorfman realised that instead of testing each person with one test each, the samples from several people could be mixed and the whole pool tested at once. If nobody in the pool had syphilis, the pooled sample would contain no syphilis infection, so the test would be negative. The army would then learn that all those soldiers were syphilis-free, and all for the cost of only one test.

On the other hand, if anyone in the pool had syphilis, their sample would give a strong enough signal that the infection could be detected, and the test would be positive. Then further investigation would be needed to work out which soldier (or soldiers) in the pool caused that positive test – Dorfman proposed simply retesting all its members individually to find out which of them had the disease.

Dorfman realised that this strategy would be particularly effective at low prevalence of the disease. In this case most pools would not contain any infected people, and so all their members could be confirmed as uninfected with a single test. Sometimes it would be necessary to do further individual testing, but this was rare enough that there were potentially considerable savings in the number of tests required using this scheme.

While Dorfman's ideas were never implemented at scale, they remained of interest to mathematicians and biologists ever since. This was often as an intellectual exercise, to design better testing strategies than Dorfman's simple scheme and to find clever ways of determining which individuals in the population were infected. However, pooled testing was also applied in problems in biology, cybersecurity and communications.

It remains an active area of research to understand the limits of what is possible in pooled testing, but Shannon's work gives one key performance bound. We want to learn as much information as possible from each test, so we want to make it roughly equally likely to be positive or negative, so that we can learn the full 1 bit per test. Additionally, we would like successive tests to be as close to independent as possible, so that the bits add up. This means that we want to mix up people thoroughly into pools that don't overlap much, rather than wasting tests on similar pools. These kinds of test strategies achieve Shannon's goals of having independent equally likely outcomes, making the tests more like the fair coin than the biased one.

A further challenge is that, as described so far, the testing is assumed to be perfect in that if a test pool contains an infected person then we will definitely receive a positive result, and otherwise we will not. This is an interesting mathematical abstraction but, as we have seen in the previous chapter, testing a sample can incur both false positive and false negative errors. This issue will only get worse as samples are mixed, and one might assume that false negatives would get more likely due to the effect of 'dilution' (one positive sample being potentially swamped by many negative ones, so that it isn't detected).

However, the problem of noise is not insurmountable from the point of view of pooled testing algorithms. The corresponding theory continues to be developed, often motivated by ideas from Information Theory, and indeed has been a major part of my own research in recent years. Partly because of all this theoretical research, pooled testing was applied at large scale in several parts of the world (including China, Israel, Rwanda and parts of the USA) during the coronavirus pandemic, allowing a valuable gain in the efficiency with which tests were used, and meaning that Shannon's ideas have had a positive impact in this sphere as well.

Shannon and filter bubbles

Another way Shannon's mathematical formulation of uncertainty and information can help us is by suggesting some principles concerning our consumption of media. Many of us have become aware of the danger of a 'filter bubble', a self-contained group of people reinforcing their own beliefs by reflecting them back to one another, but it is interesting to think that this can be formulated in a mathematical way.

We can think of the problem from the point of view of a news consumer. Sitting at home with an Internet connection, I have access to an enormous variety of sources of information regarding the state of the world. I need to choose which ones to subscribe to, and how to combine their messages into a single opinion in terms of how things are going. This can be a bewildering challenge, often referred to as 'drinking from the firehose', but mathematics can give me some advice on how to do this.

Firstly, as described previously, I should not be aiming to come up with a definitive mental picture of the state of things. Even in the best case, there will still be some uncertainty, and so I should be aiming to come up with a range of possible scenarios that might be unfolding, and to put some amount of weight on how seriously I take each of them. In the language of Chapter 6, I should be aiming for a confidence interval not a point estimate. In an ideal world, I would like my range of possibilities to contain all reasonable scenarios, but not keep an open mind on possibilities that can sensibly be dismissed.

Secondly, I don't need to understand everything. For example, suppose I am interested in energy policy and security. We would like a balanced collection of power sources which are reliable year-round and protected from international political fluctuations, while minimising CO_2 emissions. There are a range of different solutions available, and it would be a major job simply to list the policies and aims from over 200 countries, let alone to process them into a coherent mental model.

Personally, since I live in the United Kingdom, I am most interested in understanding the right plan for that country. This is not to say that other countries can simply be ignored, but in Shannon's language we should be looking for communication channels that are not *too* noisy, where information from one country might be relevant for us.

For example, while it is interesting to be aware of the solutions pursued by countries with reliable year-round sunshine, too great a reliance on solar power is unlikely to be a realistic option for the UK. Instead, Northern European countries with a similar climate may be a better comparator, and it would be worth examining what strategies they are each pursuing.

Additionally, as described above, Shannon showed us that we can learn the most from a collection of information sources that are independent of one another. While the Dutch experience may be interesting to examine, if Belgium and Luxembourg pursued similar government policies then there would be little extra to be gained from adding them into consideration as well. We would learn more from considering countries whose strategies are different. This may include France who generate over 70% of their electricity from nuclear power, or Norway who overwhelmingly use hydroelectric power stations, even if these options may not carry over exactly to the UK.

Indeed, this principle applies to our choice of information sources in general. If your aim is to gain as much information as possible from the things that you read, then Shannon shows that there is little point in even opening an article by a columnist who is so predictable that you know what they will say in advance. The entropy of their contribution is close to zero, and you will learn little added information from them. In the same way, you should not be afraid to skim read – if an article is well structured, you should learn more per word from the early paragraphs than will be added by the later ones.

Similarly, you might like to think about what is gained by following someone on Twitter whose views never change or are identical to many people you already follow. From the point of view of Shannon's Information Theory, you will learn little new from this person. Instead, you should be looking for pundits who are interesting. This does not mean that they are always right – very few people can claim that – but if you can find a collection of people who think genuinely independently of one another and who provide relevant expertise and correct insights a large enough proportion of the time, you are likely to be exposed to a range of opinions and perspectives that you would otherwise have missed. Given some judgement you can sort through these views and decide for yourself who is right on this occasion.

Indeed, mathematical language allows us to understand the dangers of a filter bubble, where you choose to only listen to the opinions of people whom you already agree with. A principle called the Wisdom of Crowds suggests that we can gain accurate estimates of unknown quantities by taking many sensible guesses and averaging them together. Just as with Fermi estimation, we can justify this principle through an appeal to the Law of Large Numbers. Assuming the guesses are formed independently of one another, if they all represent random fluctuations from the right answer, then these random errors tend to cancel out on averaging.

However, imagine a situation where the crowd does not make genuinely independent guesses. If within a crowd of 100 people, 99 of them base their opinions on the views of a particular pundit, then the weighted average of the 100 opinions will tend to be extremely close to that person's view. We no longer gain the Wisdom of Crowds, although we can fool ourselves into thinking that we do. In fact, we should give larger weighting to the one independent person to ensure that the average has the best chance of being close to the truth.

Information is power

If you do inadvertently find yourself in this kind of position, the consequences can be extremely serious. By ending up in a situation where your feed places undue weight on the opinions of one person or group, you can find yourself over-reacting to news (since one person is more likely to take an extreme view in one direction or another, compared with considering a plurality of opinions) and be prone to misinformation or distortion filtered through a narrow focus.

Further, by crudely weighting the apparent popularity of opinions based on how often you see them in your feed, you can end up with an unbalanced view of how the world outside social media thinks about an issue. If you deliberately choose to only follow people who strongly believe in Brexit or people who think that more European integration is the answer, you cannot then judge the progress of the wider argument from the opinions you read. It is right to think that 'Twitter is not the real world', and it is worth considering the plurality of your sources when judging how popular a particular policy may prove to be.

While this may be a crude model of the true situation, it is certainly worth considering whether you have engineered yourself a social media feed that is unduly influenced by a few people's views, or whether you are exposed to a true plurality of opinions. Certainly, mavericks like Shannon show the benefits of trying to think in an original way, and of trying to listen to as many views as possible before making up your own mind.

Information Theory and gambling

There is another particularly interesting connection between information and the topics of logarithms and gambling that we have already seen in Chapter 8. We have already seen how Claude Shannon's work captured the idea of information in data. One of Shannon's colleagues, John Kelly Jr., realised that the same ideas could be used to study gambling. He developed a betting strategy based on Information Theory, now known as the Kelly Criterion, which seeks to maximise the doubling rate of our fortune – that is to make the line on a logarithmic plot of our money as steep as possible.

Kelly's strategy was based on the difference between the book-makers' odds and the true underlying probabilities of outcomes. He showed that, whatever the odds, you should divide your stake between all the possible outcomes, proportionally to the underlying probabilities. It turns out that the Kelly Criterion can be too risky – although it has high expected value, it also has high variance – so often gamblers will implement a more cautious version of it.

In fact, there is a link between the pay-off you expect to see from operating the Kelly Criterion and the entropy itself. Remember that Shannon introduced the notion of entropy to quantify the idea that some random quantities are more random than others – so the fair coin had entropy of 1 bit and the biased coin (with probability 0.89 of coming up Heads) had entropy of 0.5 bits.

Imagine a very generous bookmaker who allows us to bet at evens on these two coins being tossed. Kelly tells us that with the fair coin we should split our money in two, bet half on Heads and half on Tails. Each time we will lose one stake and double our other stake, and so end up where we started. However, with the biased coin we should bet 89% of our stake on Heads and 11% of our stake on Tails each time. Since most of the time the coin will be Heads, we will typically win more than we lose, and in fact our fortune will tend to grow exponentially, at a rate that can be expressed in terms of the entropy.

This idea holds in general. The smaller the entropy, the less random the outcome is, the larger the profit we can make from betting on it at these odds. This led to a gambling game invented by Cover and King in 1978, where we can estimate the entropy by seeing how much profit we make in this scenario. Cover and King used this to estimate the entropy of the English language – in other words how unpredictable it is – by getting people to bet on the next letter that appeared in a text.

Of course, it's not realistic to expect a bookmaker to offer you odds that allow you to make a profit in this way. But occasionally casino games such as blackjack may develop so that the odds are temporarily in your favour, and if so then Kelly tells us how to bet. Once again, the interplay between odds and logarithms offers us an insight into real world problems, just as it did with the understanding of errors in medical trials above.

Summary

In this chapter, we have seen how the ideas of the genius Claude Shannon can make sense of the modern world. Shannon's entropy and its unit, the bit, are very much the currency of information and of how we live today. Shannon allows us to understand the fundamental limits of questions such as data compression and communicating in the presence of noise, and to quantify the overlap and redundancy between various sources of information. While originally introduced by Shannon to understand communication on copper telegraph wires, these ideas have flourished and spread and now give us vital insights into settings as diverse as pooled testing for disease, filter bubbles and gambling strategies.

Suggestions

You might like to explore these ideas further by thinking about the units of bits. When you look at download speeds, the size of memory cards and mobile phone contracts, you can carry out the same kind of approximate calculations that I espoused in Chapter 2 – perhaps in units of CD-ROMs (around 700 megabytes) if you are of the right generation to have used them. You might also like to think about the correlation between the sources of information that you consume, and perhaps take steps to address this, for example by buying a different newspaper to normal or following people on social media with whom you often disagree.

Chapter 10

Drunkards, queues and networks

Nuggs and memory

On 6 April 2017, an American teenager named Carter Wilkerson took to Twitter, where he had 138 followers, to ask one of the great questions of all time: 'Yo @Wendys how many retweets for a year of free chicken nuggets?' The restaurant chain Wendy's replied with '18 million', and it appeared that was that.

However, something bizarre and crazy happened. Wilkerson launched a campaign under the hashtag #NuggsForCarter, which was picked up and promoted by celebrities. It seemed like a feel-good story, where anyone could join in and retweet. Each retweet made the story more compelling, and helped his campaign gather momentum. He didn't manage 18 million, but he didn't do badly – he was retweeted at least 3.4 million times, breaking the world record previously held by the 2014 Oscars selfie posted by Ellen DeGeneres, and Wendy's were happy enough with the free publicity to give the man his nuggets.

But what has all this got to do with maths? It turns out that we can think mathematically about the way things spread, whether that be a virus in a city or a campaign on a social network.

We have already seen the value of randomness as a tool to understand the world. However, we have mostly focused on collections of numbers which are independent, rather like a sequence of coin tosses. While this is often a helpful assumption which justifies the use of results such as the Law of Large Numbers and the Central Limit Theorem, in general it may be too restrictive to capture many problems of interest.

I described in Chapter 5 how this independence arises from objects such as coins and lottery balls having no memory of the previous results. In some sense, our information about the past gets thrown away and does not affect future outcomes for these objects. However, this is really a special case. In general, there is a wide class of processes for which the information about their past behaviour affects where they go next. This can be harder to study mathematically, but the resulting behaviour can be richer and more interesting, so the effort of this study is often repaid with insights into real-world scenarios.

A classic example of this kind is a *random walk*, or drunkard's walk. Consider someone moving in a random direction along a long straight road, for example under the influence of alcohol. A simple model is to imagine that once a minute they toss a fair coin and move one step forward for Heads and one step backwards for Tails.

We can think a little bit about the way in which independence manifests itself through this model. The walker's successive coin tosses are independent, but their positions at successive times are not. If at some stage the walker has moved 50 steps forward, then a few minutes afterwards they are likely to be somewhere close to this position. Certainly, they cannot return to their starting point in less than 50 minutes. In other words, information about the walker's current position gives us a clue about where they will be in the future. However, the value of this information decays as we try to make predictions further and further forward in time.

The position reflects a partial memory of the coin tosses. If the entire sequence of tosses were repeated in an identical way, the walker would end up in the same place. However, we don't need to know the whole previous sequence of tosses to know where they end up, only how many Heads and Tails there were. Given knowledge of their current position, the only thing that we need to know to be sure where they go next is the result of the coin toss.

This is an example of a *Markov chain*, named after the Russian mathematician Andrey Markov, who formalised their definition early in the 20th century. Markov chains are a class of random models with a limited amount of memory in this sense, where although the current position influences the next position, each jump is independent of the last.

Such models are extremely useful in a variety of contexts, the classic example being finance. If you examine plots of stock prices, exchange rates or other financial quantities, you will often observe a characteristic 'wiggly' shape, as we already saw with the Dow Jones data in Chapter 3. These tracks seem to constantly be moving up and down, often reversing direction very frequently, but with a longer-term general drift in a certain direction.

Sterling vs dollar exchange rate 2022

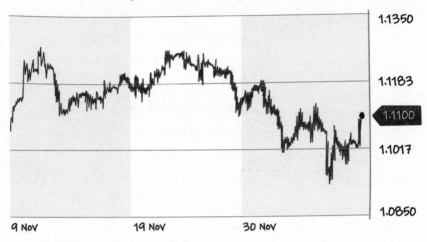

Further, whether you look at stock prices across a day or across a year, you will observe a similar kind of wiggly structure. Mathematicians refer to this as a scale-invariant property, which suggests that stock prices are a fractal, just like the brightly coloured Mandelbrot sets that seemed ubiquitous for graphic designers of the 1980s and 1990s.

Curiously, such wiggly scale-invariant behaviour is also exhibited by an abstract mathematical object known as Brownian motion, which lies in the same class of processes as Markov chains. This fact has led to Brownian motion being used to model stock prices and underlies the famous Black-Scholes formula which gives the fair price of derivative securities such as put and call options. The holder of a put option has the right, but not the obligation, to buy a stock at an agreed price and time in the future, and a call option gives the right to sell a stock on similar terms. Essentially, these options act as insurance against large price fluctuations in one direction or another. The Black-Scholes formula, for which the 1997 Nobel Prize in Economics was awarded, allows us to quantify how much this protection is worth, somewhat like knowing how to set a fair insurance premium.

However, it is important to stress that although Brownian motion is a good model for stock prices, it does not help us predict their short-term evolution in a way that would allow us to make money. Essentially, the model suggests that the prices at a future time will be consistent with a particular bell-shaped curve from Chapter 5, but it doesn't tell us *where* they will end up in the range of plausible values.

In fact, as a log scale enthusiast, it would be remiss of me not to mention that Brownian motion really models the *logarithm* of stock prices, not the prices themselves. This should not be a surprise. Remember that in Chapter 3 we saw that the value of money growing according to compound interest makes exponential growth be a natural model, and so it is often preferable to plot financial data on a log scale. This is also the right way to think about the return on your investment – you will make as much money from £1,000 invested in shares that rise from 1p to 2p as from those that rise from £10 to £20. What really matters is the multiplicative factor of increase or decrease, meaning that a log scale is the right measure for stock prices. Many finance websites provide this option for their charts.

Random walks on networks

In fact, rather than our drunkard wandering randomly on a simple straight line, we can think of a random walk exploring a more interesting region, via a more complicated set of moves. For example, consider a knight moving at random on an empty chessboard. A knight's move is an L-shaped pattern, made up of two vertical steps and one horizontal step, or two horizontal steps and one vertical step. When in the middle of the board, the piece can move to one of eight possible squares, whereas when positioned at the edges or corners it will have fewer available moves.

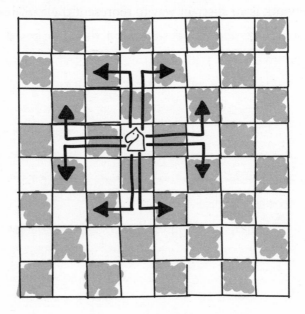

Suppose that at each time, the knight lists the possible moves it could make and chooses randomly between them with each move being equally likely. We can imagine tracking the knight's moves and studying where the piece spends time in the long run – for example by plotting a 'heat map' that shows the proportion of time spent on each square, as sometimes displayed for footballers.

This is another example of a Markov chain. As with the drunkard's walk, the only information needed to find the knight's next position is its current position and the outcome of a random choice of jumps. To understand its progress, we can introduce another mathematical object. Formally, mathematicians would call it a *graph*, but since I have been using that word to talk about two-dimensional plots of data, I will use the word *network* instead. A network is a collection of points known as *vertices*, joined by lines known as *edges*.

The network is an abstract way to represent the chessboard. The vertices each correspond to a square on the chessboard, and the edges connect pairs of vertices that a knight could legitimately jump between. Using a full eight-by-eight chessboard would give a network with 64 vertices and a tangle of edges (we'll see this picture later), so for illustration here is the same picture for a three-by-three chessboard. I have labelled the vertices with the co-ordinates of the squares of the chessboard. You can check that any legal knight's move appears as an edge of the network.

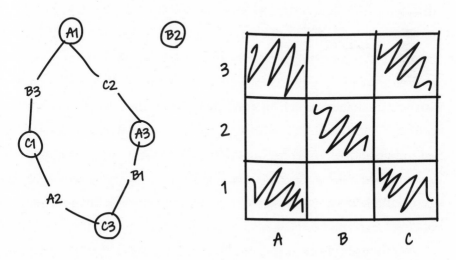

You will notice that this representation of the three-by-three chessboard unfolds it into a ring of eight vertices, accompanied by the vertex representing the central square B2 which sits isolated (since there are no legal moves to or from this square). I have marked the black squares with a circle, to emphasise the fact that each knight's move takes it from a white square to a black square or vice versa.

This representation makes clear the choice that must be made by the knight randomly walking on a three-by-three chessboard. It can make a clockwise or anti-clockwise step around the ring with each move, essentially by tossing a fair coin to decide which (say Heads for clockwise, Tails for anti-clockwise). If you know where the knight is now, the only information you need to know where it will be next is the outcome of the coin toss – how it got there doesn't matter. If you want to know where it will be in ten moves time, you need the outcome of the ten corresponding coin tosses.

It is worth remarking that because a knight can move in either direction, the edges of this network are *undirected* (can be followed in either direction). For a pawn, which can only move forward, the edges of the corresponding network would be *directed* (can only be followed in one direction, which we would represent with an arrow). Generally, much interest focuses on undirected networks of this kind, though it is worth noting that directed networks play an important role in some modelling situations. Indeed, we have already seen in Chapter 2 the role that the asymmetry between outgoing and incoming messages plays in understanding the dynamics of email transmission.

Another simple example of a directed network is Twitter, where the relationship 'X follows Y' is not necessarily mutual, in that 'Y follows X' does not necessarily happen as a result. An extreme case was Donald Trump who, before his @RealDonaldTrump account was deleted, followed 51 accounts and was followed by nearly 90 million people. We can think of information spreading across the network and see that this kind of imbalance produces considerable asymmetries in terms of how messages propagate.

If I had wanted to get some information to Donald Trump, one of my tweets would have had to be retweeted by someone who followed me. It would then have to have been retweeted in turn through the network until it eventually reached one of those 51 accounts, who retweeted it at a time when Trump might have seen it. Whereas in contrast, if Trump sent a message, since I followed many people who were likely to retweet it, it would almost unavoidably have passed in front of me, if only with a rude comment attached to it.

Returning to the three-by-three chessboard, if a knight following the rules of the random walk begins on one of the eight squares around the ring, its heat map will gradually become uniform across the squares. It is clear there will be some pattern to where the knight will be found, since they will alternate between white and black squares, but on average this won't make any difference.

In other words, on average the knight will eventually spend the same proportion of time on each of the eight vertices. This can be proved formally but is essentially a consequence of the symmetry of the picture – after a few hundred moves, the knight will lose memory of where it started and be roughly equally likely to be in any state. It's a lot like the Law of Large Numbers from Chapter 5, the randomness of the coin tosses gradually averages out.

Curiously, this can be understood in terms of Claude Shannon's entropy quantity from Chapter 9. Being equally likely to be in each state is the configuration which maximises the entropy, just as the fair coin did – these are the sets of probabilities which are most unpredictable. It turns out that there is a good reason for this. The information-theoretic quantity called entropy tends to increase as we introduce more randomness into a system, just as the related quantity called entropy in thermodynamics tends to increase over time (the Second Law of Thermodynamics). In that sense, perhaps surprisingly, the knight spending equal time in each state in the long run is the equivalent of the Central Limit Theorem that we saw in Chapter 5.[9]

In fact, a similar formula holds for someone performing a random walk on any undirected network of this kind. We call the *degree* of a vertex the number of edges coming in (or out!) of it. In the three-by-three chessboard example, there are eight vertices, each of which have degree equal to two. It turns out that the long-run proportion of time that a random walk spends on any vertex is proportional to the degree of that vertex. (There is a neat proof involving filming the moves and playing the tape in reverse, but it's a bit too complicated to explain here.) Since each edge has two ends, the total of all the degrees is twice the total number of edges of the network. This means that the long-run proportion of time a walker spends on a particular vertex is the degree of that vertex divided by twice this total number of edges.

On the standard eight-by-eight chessboard, since the 16 squares in the centre each have eight possible knight moves leaving them

---

9 These problems were the topic of my very first published mathematical research papers, but unfortunately are a bit of a digression from the main theme of this chapter!

and the squares in the corners each only have two, a random walk will spend about four times as long on a central square than on a corner. This shows that even following a relatively simple rule, the long-term behaviour of the walk can become interesting and non-uniform – some (well-connected) squares are more likely to be visited than other (less well-connected) squares.

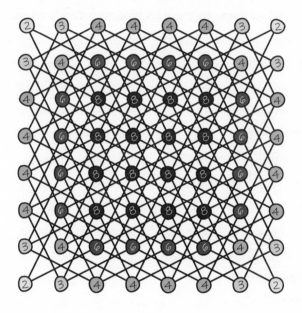

This idea of random walks spending a long time in well-connected vertices lies at the heart of the PageRank algorithm, which was published in 1998 and powered the original Google search engine. PageRank was developed by Larry Page and Sergey Brin, who thought about a web surfer exploring the Internet by following random links. Just as with our randomly moving knight, the surfer would spend more time on well-connected, high-value vertices. In fact, these are not simply pages with many incoming links, but rather those which have incoming links from other highly valued pages – being connected by the BBC is worth much more than a link from a spammer's web page, for example. Page and Brin realised that the connectivity of the Internet represented an implicit rating of trustworthiness and knew to direct searches to the highly rated sites. There has a been a constant battle ever since between Google and those seeking to manipulate their algorithm through search engine optimisation tricks, but these mathematical random walk ideas remain at the heart of Google Search.

Chessboards are well connected – even in the worst case of starting from one corner, a knight can always reach any other square of an eight-by-eight chessboard in six moves. Mathematicians would say that the *diameter* of the network is six, in analogy with the diameter of the circle, this being the longest distance between any two points. Similarly, by drawing the three-by-three chessboard as a ring of vertices we can see that it is always possible to reach any point from any other point in no more than four moves, so the diameter of that network is four.

The idea of measuring distance by counting steps across a network arises in the 'Six Degrees of Kevin Bacon' game, where someone who appeared on screen with the actor Kevin Bacon has a Bacon number of 1, someone who acted with someone with a Bacon number of 1 has a Bacon number of 2 and so on. Similar ideas occur in other contexts, such as the Morphy number which forms connections based on having played games of chess together and the Erdős number which counts jointly published mathematical papers.[10]

---

10 For what it's worth, my own Erdős number is 3, my Morphy number is probably very high, and my Bacon number may be 4 depending which technicalities you are prepared to accept.

Drunkards, queues and networks

In each case, the number is based around finding shortest paths to an interesting vertex, whether it be Kevin Bacon, 19th-century chess champion Paul Morphy or hyper-prolific Hungarian mathematician Paul Erdős. In that sense, the informal claim that all actors have a Bacon number of 6 or less is not quite the same as saying that the diameter is 6. There could well be two actors who each have a Bacon number of 6, but with the shortest path between them going through Bacon himself, in which case we would need 12 steps to connect them. However, if every actor's Bacon number is no higher than 6, then the diameter can be no more than 12.

However, an interesting question is what happens if certain vertices are unavailable – we could think of cutting squares out of a chessboard, or a piece of the same colour already occupying it. In an abstract model of a computer network, this captures the idea that one machine may fail and that information may need to be rerouted to avoid this. For example, in the three-by-three case, if we want to reach A1 from A3, our ring diagram shows that it takes two moves. However, if the square C2 is removed, we have no choice but to go the long way around, meaning that we require six moves.

Removing a small number of vertices from a network can have a major impact on how efficiently it is connected. For example, if Kevin Bacon and all his movies had never existed, it is likely that many of the shortest paths between pairs of actors would become significantly longer. This loss of connectivity would be unwelcome when considering a computer network where we wanted to spread information efficiently. On the other hand, we will see shortly that random walks can model the spread of an epidemic, with vertices corresponding to people. In that sense, removing vertices so that the epidemic cannot spread through them (by vaccination, for example) would be *good* news.

In either case, it is interesting for mathematicians to study how resilient a network is to failures of this kind. They have developed specific measures to capture whether a network has 'bottlenecks' through which many paths are routed, including a measure of resilience known as the Cheeger constant. The Cheeger constant of a network can tell us how fast information or an epidemic would spread on it, and how fast the heat map gets close to its final averaged state.

While I have described a setting where a random walk is equally likely to make any possible move, there are more general Markov chain models on networks. These essentially correspond to tossing a biased coin to decide where to move next. In fact, much of the same theory can be used in this case to understand what the long-run heat map of the walker looks like.

Queueing networks

One further interesting type of Markov chain is known as a queueing network. The mathematical study of such networks was started by a Danish researcher called Agner Krarup Erlang early in the 20th century to understand the properties of the Copenhagen telephone exchange. In fact, we can think of the simplest version of this model, consisting of a single queue of people waiting for service, for example at a post office till. This can model service processes of many kinds, including the service of data packets over the Internet through a router.

The length of the queue changes in two ways. Firstly, customers arrive at random times and join the back of the queue. Secondly, customers are served one by one at the front of the queue on a first come, first served basis, and the amount of time this takes (referred to as the *service time*) is random. Interestingly, just like the drunkard's walk, this forms a Markov chain: to know how long the queue will be next the only information we need is its current length and whether anyone has arrived or been served in the last time instant.

However, even this simplest model has some interesting features. If on average customers arrive faster than they are served, then the queue length grows out of control, so we tend to ignore that case. However, even when the average arrival rate is less than the service rate, the queue can grow arbitrarily long for a while.

We can think about a heat map of queue length – what proportion of time is spent with nobody in the queue, with one person in the queue and so on. It turns out that there is an exact expression for this long-run proportion, which is given by something called a geometric distribution. This shows that the proportion of time that the queue has more than L people in it decays exponentially as L gets large. However, the queue could temporarily grow very long indeed, until the service process reduces it down to a more manageable size.

There are an enormous number of variations of this model, where the randomness of arrival and service time is specified in a variety of ways, there are multiple servers, the queueing discipline is different, customers can drop out of the queue, or where having been served in one queue the customer is routed through a network like the Internet. However, even this simple model is interesting in the context of epidemics, particularly when considering healthcare provision.

Understanding COVID and economic numbers

Having seen that a random walk evolves with some degree of memory rather than in a completely independent way, this theory can help us understand various aspects of epidemic spread and other settings. One obvious place where random walks were relevant was the daily reported COVID numbers, which behaved with a degree of memory in exactly this sense.

In essence, the reason for this is that the number of infected people evolved gradually. We can think of each person being infected for a period of, say, ten days. The exact time does not matter too much here and may depend on exactly what definition is used; whether it is the time spent showing symptoms, being infectious, testing positive with a PCR test, or something else.

However, roughly speaking the number of infected people on one day will be reasonably close to the number of infected people the day before. Some new people will become infected, some infected people will recover, but for people who were infected between one and nine days ago nothing will change. On average, this might mean that 90% of the infected population stays the same on a day-to-day basis, and the only change comes according to whether more people get infected than recover. This entire process acts somewhat like the queue model that I described above.

Now, as I have described before, we never got to directly measure the number of infected people at any given time. However, the numbers reported, such as cases, hospitalisations and deaths, depended on the size of the infected population, albeit perhaps with lag from the time of infection. Since a certain proportion of infected people tested positive, go to hospital or die, and in the short term this proportion probably did not change too much, the daily numbers depended on the size of the infected population in a somewhat predictable way.

Hence, because the overall infected numbers wouldn't change too much from day to day, neither did the announced numbers, so a sensible first guess is that each day's numbers would have been somewhat like the previous day's numbers. I will discuss other fluctuations in the next chapter, including some that depend on the day of the week, but for now this is a surprisingly good first model. Just like the drunkard's walk, knowing current values of these figures gives good information about future values, at least in the short term.

This phenomenon can be observed in other time series, for example in reported economic data. We can imagine that, like the number of infections, there are underlying variables that roughly summarise the health of the economy. For the most part they probably evolve slowly, with the rare exception of severe and unexpected shocks such as the 9/11 attacks.

We cannot observe these underlying variables directly, but we see aspects of them through reported numbers such as unemployment, inflation and growth figures. But it is always worth remembering that the reported numbers are imperfect in the same way as the COVID numbers were, in that they are noisy and lagging snapshots of an underlying process that we really hope to monitor. However, the presence of uncertainty and noise in the reported data does not mean that it is worthless. Rather, we need to bear uncertainty in mind and try to view the whole picture instead of jumping to conclusions from one piece of data.

Going viral: spread on a network

Models of random walks on networks can be extremely helpful to understand how exactly information or infection spreads. If we think about social media connectivity, we could imagine Twitter users as vertices on a network and connect pairs of vertices with edges if one follows the other (as described before, we treat this as a directed network, with arrows indicating who follows whom). We could visualise Carter Wilkerson's infamous chicken nuggets tweet 'going viral' on this network – it would start at some vertex, spread out along edges (following the direction of the arrows) to reach new vertices, from which it can be retweeted to further vertices and so on.

Indeed, much of the new mechanism of Twitter is designed to make certain tweets go viral in this way – once a tweet has achieved a certain amount of success, it may get added into lists of trending topics and so on. However, certain people will have a disproportionate effect in determining how successfully a particular tweet spreads: if someone with a million or more subscribers retweets it, then instantly it will be seen by a lot of people, many of whom will retweet it in turn. A binary decision taken by one person deciding whether to press the retweet button can therefore have a huge effect. It's very possible that in a parallel universe almost nobody saw Wilkerson's tweet, and that he never got his free chicken nuggets.

It is no coincidence that the phrase 'going viral' describes a message which successfully spreads on social media. Indeed, we can model the spread of infection between individuals in precisely the same way as this spread of information on a social network. While everyday connections between people are not quite so well defined and easy to map out as those between Twitter followers, we can imagine the network of a human population in an analogous way. Again, individuals form the vertices of a network and we connect pairs of people with an edge if they are a frequent contact of one another. In this case, it probably makes sense to think of this as an undirected network, since face-to-face encounters are reciprocal in a way that following someone on Twitter is not.

We can think of a viral disease spreading across this network somewhat like the knight performing the random walk on the chessboard. Starting with Patient Zero, whoever they may have been, the virus would move between social connections at random. One difference from the knight model is that one person can infect multiple people; technically this is referred to as a *branching random walk*. However, many of the same features should still hold.

For example, people will have varying numbers of contacts, and those with the most contacts are likely to have a disproportionate effect on the overall rate of spread of the virus (such people are often referred to as 'super-spreaders'). Just as the knight spent more time on a square that was well connected, such people are more likely to be infected early on.

In fact, the network would constantly evolve as infections took place. Assuming that people could not be infected twice, once a person became immune, that vertex would act as a roadblock to the spread of the virus. Since as described the vertices with higher degree (people with more connections) are more likely to be infected early on, and their removal from the network would cut more paths, this gives some optimism that the Herd Immunity Threshold could be lower than conventional epidemic models suggest.

However, there may be limits to the accuracy of treating the spread of the virus purely as a uniform random walk on a network. First, it seems likely that some pairs of contacts will be more important than others – you form a more consequential connection with the people you share a house with than the person you buy coffee from.

To some extent, the basic mathematical model can be adapted to cope with this: instead of simply connecting people with an edge, we could label the edge with a number reflecting the frequency of contact, and assume the virus is more likely to spread along edges with a bigger number. Equivalently, the virus has different probabilities of moving along different connections.

However, it may be unrealistic to think that most people have a static network of contacts. While they may have a core set of people (housemates, friends, colleagues) they see on a regular basis, this is likely to be augmented by an ever-varying set of other people (bar staff, bus drivers, people in the supermarket queue). Many of these variable contacts would be on a one-off basis, and to capture that in a network would be fiendishly complicated. Hence, the abstract mathematical model may be too simplistic to entirely explain the spread of the virus, though it can certainly help our understanding.

Queue for hospital resources

We can think of the allocation of hospital resources using the queueing network models I described above. For example, we can think of the allocation of patients to beds as being somewhat like the post office queue, but with multiple cashiers serving customers.

Each bed corresponds to a cashier in this sense – a patient requires it for a variable period, until they can be discharged. Patients each arrive at a random time. Ideally, we want to keep some beds free, so that whenever a patient requires one, then it is immediately available for them. Of course, there will be additional constraints, since not all beds are the same – there is little point having a bed available in Cornwall if a patient is in Aberdeen, and the bed requires resources and staff to ensure that it operates effectively. However, this works well as a simple first model.

In an idealised situation, the time that each patient spends in the bed would be random, but the typical lengths of stay would change little from day to day. That is, we would hope that there would be enough beds available that a patient can occupy them until they are 'cured', unaffected by the level of demand. In practice, we may be able to somewhat shorten the time each patient spends in hospital if the beds are under pressure, but there probably remains a minimum necessary length of stay for each one. Hence, it is reasonable to think that a random process determines the time each patient occupies the bed, referred to as the service time in the queueing model.

Making this assumption, we can think about how the queue behaves as the number of patients changes. Essentially, as demand for healthcare grows, patients would eventually start to arrive faster than they are being discharged. In a scenario where the epidemic is growing exponentially, demand inevitably outstrips supply at some stage, meaning that at some stage beds would no longer be available and the length of the queue would start to grow. Unfortunately, this would mean that some patients would suffer health consequences or even death resulting from a lack of treatment.

However, just as the queue could start to grow in the simple case of a single server described previously, in a worst-case scenario random fluctuations could still lead to a temporary lack of resources, even if the rate of arrivals was less than the average rate of discharges. For example, a lot of patients in one hospital could all randomly have unexpectedly long hospital stays at the same time. In total, across the healthcare system, the Law of Large Numbers suggests that these effects will average themselves out. However, if we are concerned with one hospital with a relatively small number of specialist beds of a certain type, then this effect can be a major concern for the management of healthcare resources.

We can understand other settings of competition for resources in this framework. Routing of computer data, both within a network of computers and between the separate processors of a PC, is governed by the same principles and is studied in the same way. One more familiar setting where queueing theory provides insights is road traffic management. We are of course all used to sitting in a traffic jam when road capacity is exceeded by the number of cars wishing to use it, matching the idea of arrival rate exceeding service rate.

However, by analysing road networks, it is possible to gain insight into phenomena such as the way in which a small drop in total traffic in school holidays can lead to noticeably increased flow (again by taking the arrival rate below a critical level). Unexpected results can arise, such as Braess's paradox, which shows that adding new roads to a network can sometimes reduce the overall rate of flow through it by creating extra demand at bottlenecks elsewhere in the network. Such effects have been observed in practice in many countries, and the relevant mathematical theory must be considered to ensure that this does not occur.

Summary

Overall, understanding randomness beyond simple independent models is a valuable tool, and processes which evolve according to information about their current position are a natural next step. These include random walks and Brownian motion, including the model of random walks on a network. These Markov processes allow us to understand better the behaviour of stock prices, daily pandemic numbers and queues, as well as the idea of viral spread on a network, whether metaphorically or literally meant.

Suggestions

You can explore these ideas further by observing the behaviour of queues next time you go to the post office or IKEA. You might like to look up tracks of stocks and see which ones seem to obey the Brownian motion model – in fact, this is only an approximation, and as discussed in Chapter 5 stock prices may fluctuate more wildly than this suggests, including the possibility of jumping to a new level when good or bad news is announced. You can think about the ideas of heat maps and varying levels of traffic by looking at which areas of carpet or lawn get particularly worn or messy – these will be the high-traffic areas, which will tend to be well connected. Can you see a noticeable effect? To think about networks,

can you find your Messi number? That is, have you played football
with someone who has played football with someone who has . . .?
Or if you are musical, what about your Sabbath number? How long
a chain of 'performed together' would you need to reach a member
of the band Black Sabbath?

Chapter 11
Measure for measure

Changing measurements

In 1998, following a nine-and-a-half-month flight from Earth, NASA's $328 million Mars Climate Orbiter mission failed catastrophically because Lockheed Martin's software had specified the impulse of its thrusters in imperial pound force-second units, whereas NASA's software expected them in metric Newton-seconds. In February 2021, journalist Liam Thorp received an early COVID vaccination appointment because his height of 6 foot 2 inches being mistakenly recorded as 6.2cm resulted in an apparent BMI of 28,000, placing him in the at-risk morbidly obese category. In 1984, legendary rock band Spinal Tap went on stage expecting to perform with an 18-foot-high model of Stonehenge, only to discover an 18-inch model had been ordered by mistake. All these stories seem too ridiculous to be true (and at least one of them may be), but they all arose from the same issue, that of mis-specified measurements. Recording and reporting measurements may not seem like a glamorous topic, but it is one where mistakes can have serious consequences, and it is worth considering further.

It is tempting to think of reported government statistics as perfect and consistent measurements, giving us comprehensive and complete information about the state of the world. However, there can be large variations in the way that things are measured over time and between different countries, and considerable seasonality effects in many datasets that mean we need to approach them with caution.

We are used to the idea that measurements are consistent. The SI units that are used for scientific experiments have been standardised to an infinitesimal degree over the years to ensure that a length of a metre or a mass of a kilogram can be referenced in every laboratory in the world. We all have measuring devices that work relative to those Platonic ideals of measurement – while a wooden metre rule in the classroom may have been knocked about over the years, we still expect that it will be roughly a metre long. If we count the number of lengths of a metre rule from end to end of a room, we don't expect an exact answer like we would obtain using high-precision laser equipment, but we might reasonably expect that the information that we gain would be close to accurate.

Further, because errors would be introduced at random, for example by not positioning the ruler exactly in the right place each time, we might hope that successive errors would partly cancel out due to the Law of Large Numbers, as they did in the example of Fermi estimation. In other words, we do not expect systematic errors.

However, if we went into the back of the school cupboard,

pulled out an old yardstick and measured the room in terms of the number of lengths of that, clearly the answer would not be correct. A yard is shorter, and so we would expect a measurement of the room based on counting lengths of the yardstick would be around 10% too large, in a systematic and consistent way.

It's worth noticing that the yardstick may not be a terrible way to measure things: for example, if we wanted to decide if a particular piece of furniture would fit in the room, then we could measure both using the same device to obtain a useful answer. However, if we wished to decide which of two rooms was bigger, then clearly measuring one with a metre rule and one with a yardstick could give us misleading information.

Further, at least the two measurements would be consistent over time. That is, if we measured the same room with the same device on successive days, we would expect to see a roughly consistent answer. We can only imagine a nightmare scenario where the length of the ruler changed from day to day, or where different rulers were used at the weekend and on weekdays. This may seem an unlikely analogy, but we shall see that something close to this may have been true for COVID data.

Further, Claude Shannon's former office mate Richard Hamming made an important point about measurements: 'There is always a tendency to grab the hard, firm measurement, though it may be quite irrelevant as compared to the soft one which in the long run may be much more relevant to your goals. Accuracy of measurement tends to get confused with relevance of measurement, much more than most people believe. That a measurement is accurate, reproducible, and easy to make does not mean it should be done, instead a much poorer one which is more closely related to your goals may be much more preferable.'

Hamming's work predated the modern era of Big Data, where vast amounts of data on every phenomenon can be easily measured using cheap sensors, reported using ubiquitous fast wireless networks, and stored and processed on modern computers. However, his conclusion remains extremely valid: just because we *can* measure something does not mean that we *should*.

It is important to find the right indicator at the right time, rather than attempt to consider all indicators simultaneously. In the Aston Villa vs Liverpool game which I discussed in Chapter 5, while it took a lot of work and analysis to find that Villa only had the ball 30% of the time, anyone who could count could see that the more important indicator was the seven goals that they scored.

In general, you should not be deceived into thinking that an irrelevant but precisely stated measurement necessarily carries any more useful information than a more relevant but approximate one. Quoting the Expected Goals figure to two decimal places may give a veneer of scientific precision, but in practice it would be almost impossible to distinguish a team who had an xG of 3.08 from one with an xG of 3.

Spurious trends

A striking feature of the experience of tracking data is our desire to impose a narrative on the numbers. For example, in the run-up to an election, people who are invested in the success of one party can often convince themselves there are patterns or trends in opinion poll data, which may just be the result of random fluctuations.

Measure for measure

A similar phenomenon is often observed in the world of finance. As I described in Chapter 10, a natural and successful model for stock prices is Brownian motion. This can be thought of as arising from a sequence of independent coin tosses, somewhat like the drunkard's walk, and is equally likely to jump up or down next, and by symmetric amounts. This means that, if stock prices really are modelled by Brownian motion, then they are essentially unpredictable. Despite this, a whole school of 'chartists' attempt to spot characteristic patterns in tracks of financial data and use them to make predictions. It is unclear that this is a successful strategy.

Part of the problem is that just as human beings are not good at generating random numbers, they are not good at deciding whether numbers are truly random. For example, if we were presented with what was claimed to be the sequence of results of tossing 200 coins, we might take a run of 7 successive Heads as good evidence that the coin was not fair, or that the results were not generated independently. However, a run of this length is exactly what you would expect by random chance – in fact, the absence of runs of 6 or 7 Heads would be more suspicious.

A similar phenomenon – where apparent coincidences are much more likely than you might expect – is the so-called birthday problem. Given 23 people in a room, there is roughly a 50% probability that two of them share the same birthday. With 40 people in a room, this rises to 90%, and with 60 people the probability of such a coincidence becomes over 99%. These may seem like surprisingly high probabilities, but the key to understand this is that there are many more chances for coincidences than you might imagine. That is, with 23 people, there are 253 pairs of people who could share a birthday, and so there are 253 possibilities for the coincidence to occur. If we consider 60 people, there are 1,740 pairs, all of which must avoid having the same birthday.

In the same way, when we toss 200 coins, there are 194 runs of length 7 to consider. Each of these has a 1 in 128 chance of containing 7 Heads, so on average we are not surprised to see a run of this length. Although overlapping runs (say starting at 1 and 3 tosses) are not independent of one another, we can analyse the probabilities more carefully and see that there is a good chance of seeing a run of this length.

In general, it is easy to convince yourself that an upwards or downwards trend is present in a plot of data, even when the numbers are generated purely at random. This is an important justification for rigorously testing whether there is really a correlation, by calculating a confidence interval for the slope of a best-fit line. If this confidence interval contains the value zero, then the most parsimonious explanation is that there is no slope at all, and that the data is essentially random with no real trend contained within it.

In the same way, data points that are distributed at random in space will tend to form clusters to some extent, because to avoid clusters the points would need to be evenly distributed in a regular pattern. Such regularity is not the hallmark of true randomness, but human judgement is not good at perceiving this.

This is sometimes referred to as the 'clustering illusion', referring to the natural tendency to erroneously deduce that apparent structure in the data has not occurred at random. A related logical error is known as the Texas Sharpshooter Fallacy, based on an old joke about someone who fires their gun at a wall and then paints a target around the points where the bullets hit. Of course, this seems ridiculous, but it illustrates the clear principle that it is necessary to formulate scientific hypotheses in advance and validate them with independently measured data, rather than to simply look for hypotheses that apparently explain the patterns observed.

Comparing countries through their data

While it was natural to try to compare the effectiveness of different governments' responses to the pandemic, this needed to be done with caution, precisely because of some of the issues of measurement discussed above. While the analogy of measuring a room with both a metre rule and a yardstick may seem somewhat contrived, there was certainly wide variation in how data was collected and what it meant.

For example, case data measured the number of people who tested positive on a certain day. However, tests were easier to obtain in some countries than in others. The ability to process many PCR tests required a well-developed laboratory infrastructure, including expensive machines and well-trained staff. Clearly this was more readily available in developed countries than developing ones, but there was even variation between European countries. Further, government policies affected the ease of getting a test: whether they were restricted to people showing symptoms, whether contacts of those who tested positive were tested and so on.

These issues mean that to compare countries simply by looking at the number of positive tests may be deceptive. However, one might assume that counting deaths gives a reasonable way of comparing the severity of the disease. Again, however, this is a minefield due to different reporting standards. Some countries counted only those deaths of people who tested positive for COVID, whereas in others even suspected COVID was counted.

There could be wide variation in timing – in the summer of 2020, the UK moved from a standard where the death of anyone who had ever tested positive for COVID would be counted as a COVID death, to two parallel standards based on needing to have tested positive within 28 and 60 days of death. This change meant that the number of reported deaths dropped overnight by 5,377 from the previously reported 42,072.

If changing the standard in the UK had such a dramatic difference (13% of the previously reported deaths no longer counting), we might assume that a sizeable proportion of the variation between countries was due to similar effects. Indeed, as with the email counting example of Chapter 2, there may well be no right or wrong way to answer these questions. It seems clear that someone admitted to hospital because of breathing problems, who tested positive on admission and died after five weeks on a ventilator should be counted as a COVID death, but whether this is true may depend on the country in question.

Further, it seems highly likely that some countries felt under political pressure to underreport deaths, since this data was being used to rate their government response. For these reasons and others, many people suggested that countries should be compared by 'Excess Deaths'. That is, if we know how many deaths normally take place in a week or year, then we can compare this with the actual number of observed deaths. This is an appealing idea, but there are several issues.

Measure for measure

Firstly, the baseline number of deaths is not a fixed and known quantity and must itself be estimated. We usually do this by taking the average of the previous five years' reported figures. However, there can be clear variation from year to year (for example, extremes of hot or cold weather in a particular week of even one year can have a major impact on this average by causing many deaths), and an ageing population can lead to increasing numbers of deaths even without a pandemic. This means that even statisticians may disagree on the interpretation of Excess Deaths figures.

Secondly, some countries report data much faster than others, meaning that early in the pandemic a comparison based on Excess Deaths could be misleading, since it may simply have been affected by reporting lags. Even within the UK, there could be variability between weeks due to registry offices being closed over Bank Holidays, if deaths were assigned to the date when they were reported, not when they took place.

Finally, we cannot simply assume that every single Excess Death during the period of the pandemic should be counted as a COVID death. While the debate about lockdowns and their impact became extreme and politicised, and the overall effect on the population's physical and mental health remains uncertain, it seems clear that there has been some impact on numbers of cancer screenings missed, for example.

If pressure on healthcare resources due to COVID led to the cancellation of screening or elective surgery, it is unclear and arguable whether the resulting fatalities should count as a COVID death or not. Certainly, they could be reported as Excess Deaths, and that should be borne in mind when making comparisons between countries based on these figures.

Variability by day

One striking fact about the coronavirus data reported around the world was the seven-day cycle associated with much of it. Roughly speaking, weekends were different to the rest of the week in a variety of ways which affected the reported data, starting with the process of infections themselves. Broadly speaking, people behave differently at the weekends than on weekdays (though when making international comparisons, it is worth bearing in mind that for a Muslim or a Jew, which days form the weekend may be different than for a Christian).

It is not necessarily that weekends or weekdays are intrinsically riskier than the other, and widespread working from home may have changed the balance in this regard. However, during the week, people were more likely to travel on public transport at crowded peak times or visit busy offices. These were activities that came with risk of infection, though equally weekend activities such as socialising or indoor exercise have similar effects, when lockdown rules allowed them to take place.

However, we did not directly observe the number of infections, but instead saw cases, hospitalisations and deaths which are 'smeared out' in time by random lags, and hence we might imagine would not show such strong day of the week effects. This would be wrong, since there are also effects in terms of how this data was measured and reported.

For example, laboratory staff being less likely to work at weekends meant that PCR tests were less likely to be processed then. People were advised not to use home PCR testing kits on Sundays because the postal service did not work then, to ensure as fresh a sample as possible was received in the lab. Later in the pandemic this pattern was reversed by a trend of Sunday lateral flow tests before the start of the school week.

Similarly, hospital bed occupancies showed a degree of weekly pattern. For example, it seems plausible that more patients are discharged on a Friday, 'to get people home for the weekend' and to slightly reduce weekend staff requirements, and that conversely physiotherapists or occupational therapists may not be available to help with discharges over the weekend. All this will contribute to a day of the week effect.

A particular effect is noticeable in terms of death data, which was presented both by day of death and by day of report. The day of death data was smooth, with no tangible strong day of the week effect, whereas the day of reporting data was usually much lower on Sundays and Mondays (probably due to fewer staff working to register deaths on Saturdays and Sundays).

All of this meant that we must approach the data with a degree of caution. Certainly, it was wrong to infer trends on a day-to-day basis, when there can be a huge degree of variability. One way that this was dealt with was to work with the seven-day average of figures, which tends to smooth out fluctuations due to these kinds of day of the week effects. An alternative method was simply to compare each day's data with the corresponding figure quoted a week ago.

However, all these techniques were somewhat ad hoc. A branch of statistics called time series analysis deals with data of this kind, using methods that can automatically detect periodic behaviour on a range of timescales. Such time series methods can often find the trends in data and beat an analysis by the human eye that can be fooled by random fluctuations.

It is worth bearing in mind that, while the mechanisms that generate it may not be so transparent, many datasets will suffer from the same kind of day of the week or time of the year effects as the coronavirus numbers. Sometimes these are explicitly taken account of, for example via seasonally adjusted unemployment numbers, but it is always worth asking if a change in a reported figure may simply be due to timing effects, and to think about how such issues might arise.

Opinion polls and surveys

Reported results of opinion polls and surveys can often seem very convincing, but it is also worth bearing in mind some caveats associated with this kind of data.

Firstly, as I described in Chapter 6, a really accurate survey requires that the underlying population be sampled randomly. This means that each person has an equal chance of being asked to participate, independently of one another. This is easier said than done.

For example, perhaps in the past we could have built a random sample of the population by picking a random entry from the phone book. However, in the 21st century a large majority of people do not appear in the phone book at all. Indeed, attempting to randomly sample by contacting landline numbers would suffer from the fact that many young people do not have a landline or live in a shared house with more adults than the national average, so would be underrepresented in an overall sample. Similarly, surveying people by grabbing them as they walk down the street would not give a representative sample, because it would potentially miss people with mobility or health issues, or who are at work when the survey is taken.

However, even if we were able to somehow contact a representative sample of people, the next problem is that not all of them would choose to answer the questions. Someone who is busy or in a hurry is less likely to respond than someone who is bored and lonely. Some polling companies attempt to overcome this by offering financial rewards for people to participate in polls – but again this is more likely to attract people for whom a small payment seems more valuable.

For all these reasons, it is safe to assume that even with the best methodology in the world, it is hard to find a truly representative sample. As a result, opinion pollsters will often try to weight their sample. If they have too few young people in their sample, they will simply count each young person's opinion more. Similarly, they may try to weight according to income or social class, using identifiers such as political party loyalty, level of education or choice of daily newspaper.

Even this weighting suffers from problems. If a young person's opinions are weighted up to balance the sample but they are not representative of their age group, this can have a distorting effect. Indeed, the problem remains, in terms of how to find a representative sample of young people. At the very least, a poll where a large amount of weighting was required will have larger uncertainty than one without, and the degree of weighting is often only visible to polling nerds who are motivated to dig through data looking for it.

On the other hand, at least the polling industry is regulated to some extent and holds itself to certain standards. For example, polling companies have agreed that the outcome of a poll carried out on behalf of a client must be reported even if the result isn't what that client might like.

However, sometimes data is reported from other polling companies, and this can bring in biases, perhaps inadvertently. If we return to the 'how much email do we get?' question of Chapter 2, if a survey had been carried out on behalf of a company selling tools to master your inbox, you might reasonably assume they would like to hear a high figure. It is possible to generate this kind of response by adding earlier questions such as 'how stressed do you normally feel at work?', and 'do you feel that your job overloads you?' to make the respondent start thinking about these issues. This is referred to as push polling, and again is generally frowned on by reputable polling companies. Nonetheless, given reports of the answer to only one question it can be hard to know if push polling has taken place.

The real Wild West of opinion surveys comes through samples which are self-selecting and self-reporting. For example, if I used my Twitter account for One Direction fans to carry out a poll asking, 'Who are the greatest boy band of all time?', we can imagine that one response might be overrepresented (at least until the BTS fandom flooded the poll). It may seem obviously nonsensical, but sometimes these kinds of self-selecting Twitter polls or surveys of the membership of organisations are reported in a way that implies that they are representative of the public.

Similar issues arise with surveys based on self-reporting of data. In the email example, if I ask people to tell me how many messages they receive every day, the result will be filtered through their perception of the situation and prone to exaggeration. In the same way, people may report the behaviour that they aspire to, rather than what they do in practice. For example, estimating UK alcohol consumption by collating the answers that patients gave to their GPs would give a rather different picture than by calculating directly from sales figures.

Wherever possible, it is better to use these dispassionate means of measurement. For example, it was probably preferable to refer to Google mobility data rather than asking people whether they adhered to lockdown. But it's important to remember that even these apparently hands-off measures can be subject to unseen biases, by only sampling smartphone owners, and so on.

For all these reasons, even if a survey has been performed and reported, it would be wise to look carefully at the details of how it was done, and why.

Data visualisation

While, as we've seen, data visualisation can be a powerful tool in communicating information about the world, there are certain principles that are worth bearing in mind when preparing graphs or presentations of your own. These will ensure you can present the data more clearly.

As a rule, people generally overestimate how clear their graph will be to a casual viewer. Having spent time preparing the visualisation and having a background understanding of the data, the graph creator already knows part of the story. It is hard not to assume that other people will share that understanding. It is important to look at the graph with as fresh eyes as possible and think if the message would still be as clear without this additional information, or to check with someone who isn't so familiar with it all.

Further, data visualisations are often prepared using sophisticated software by someone sitting close to large high-definition monitors. This means that they are tempted to add labels in small fonts to axes or points, or to use overly similar colours or different small symbols to distinguish different data series. However, graphs that you tweet may well be read by someone on a small phone screen in bad lighting conditions. Similarly, the graphs that you include in a PowerPoint presentation will be seen by people sitting at the back of the room, perhaps with others in the way and with outside light shining in. Further, remember that a reasonable proportion of the population are visually impaired to some degree, including sufferers of red-green colour blindness. It is important to try to look at your graphs in a variety of non-ideal conditions bearing in mind these issues, and to not rely on fancy plotting tricks.

Indeed, in general, I strongly believe that less is more when it comes to data visualisations. Just because it is possible to plot very many different sources of data on the same graph does not necessarily mean that it is a good idea to do so. In my view the graphs of hospitalisations in the North West that I described in Chapter 6 worked effectively precisely because they were stark and simple. I could have added more UK regions for comparison, each with their own previous level of capacity, and colour-coded them all separately. However, in the spirit of Shannon you should ask what extra information would have been added by doing that. We would expect all regions to behave roughly similarly, so having several tracks doing the same may have just provided more clutter, at the cost of little extra explanatory power.

One final point to consider when presenting data via graphs is that, while they should speak for themselves, they don't necessarily have to. When you put up a slide in a presentation, or share a graph via social media, it can be helpful to narrate it to give extra information. Say what the quantities on the x-axis and the y-axis are, and even say what a good outcome is: Do we want the points to be sloping up or sloping down? How does what we see compare to that? Are we pleased with the graph?

Overall, I believe that representing information via graphs is a valuable and compelling way to communicate, but it can be done much better by thinking about issues like these.

Summary

In this chapter we have seen that while it is tempting to make comparisons between data from different sources, there can be subtleties that make this challenging. Even within a single stream of data, such as daily coronavirus case data from one country, there can be short-term day of the week effects, or one-off changes caused by adopting a new definition or data standard. All of this means that comparisons and measurements in general should be approached with care, because they can be used to mislead. I have also discussed ways in which opinion polls and surveys can be helpful, but suggested things to watch out for, and provided some suggestions for successful data visualisation.

Suggestions

You might like to put these ideas into action in your own presentations, and to think more about which data visualisations you find convincing, and why. Next time you see a surprising opinion poll, you might like to see if you can find the details via the company's website and see if you can spot anything surprising about the sample (do there seem to be disproportionate numbers of older or younger people, of Remainers or Leavers?). Similarly, next time you hear a surprising change in data (a big year-on-year increase or decrease, for example), you might like to check whether reporting standards have changed, or whether the data might not be directly comparable to previously for some other reason.

Chapter 12
Game theory

Solid like a rock

In 2006 an old sport made its way to the big time. The 257 competitors who had qualified for the national championships travelled to Las Vegas where, in a televised competition, the winner took home a prize of $50,000. Tournaments have been played all over the world, with clips on YouTube and deep analysis of strategy taking place online. Indeed, many teams of programmers have sought to develop the perfect computer player, which can outperform both its human rivals and other competing algorithms. The sport? Rock Paper Scissors.

This may seem like a joke, but it exemplifies the importance of two of the themes of this book, namely information and randomness. It turns out that there is an unbeatable Rock Paper Scissors strategy, to play each move independently at random, making each move with probability 1/3. We can think that the other player decides their move first. A third of the time the random player will draw, a third of the time they will win, and a third of the time they will lose, regardless of what the other player does. As a result, the expected value will be zero, and in the long run the random player will break even.

However, as we have seen, human beings are bad at thinking about and creating randomness and tend to follow strategies which betray some of that. For example, people don't like repeating a move (whereas, really, we should repeat a third of the time). Hence if a player just played Rock, they may be less likely to play Rock again, meaning that Scissors becomes a safer move with the potential to win. If we have any information at all about our opponent's next move, we can exploit that, and do better than the random strategy. Hence, between human players, Rock Paper Scissors becomes a game of double bluff across repeated rounds, seeking to exploit the information that is present.

Many of the situations I have described so far have been very static, involving just one system behaving in a way that doesn't interact with others. For example, whether *my* test is a false positive does not affect the outcome of *your* test. However, the real world is more complicated than this, with interactions constantly going on between people, just like the Rock Paper Scissors game. If I choose to behave in a certain way, it may have consequences for you and affect your choices, which in turn has consequences for me.

Many of the policy issues arising from the pandemic were difficult for precisely this reason. For example, lockdowns were imposed as a policy to protect the old and physically vulnerable, but often had a disproportionate effect on the young and economically weak. Indeed, the question of whether individuals should adhere to lockdown restrictions or take a vaccination to help protect the entire community was often a heated one. There were accusations of selfishness and irrational behaviour, and sometimes an uncomfortable amount of heated inter-generational rhetoric.

It is possible to understand some of these questions in a mathematical framework, again often via toy problems. In fact, since we have been talking about toy problems and toy models, it is perhaps appropriate that this area of research is referred to as game theory. The field was really started by the work of John von Neumann (he of the 'elephant waving its trunk' in Chapter 1) in the 1920s. It rose to prominence in the 1950s and beyond as it became used as a model in the Cold War for questions such as Mutually Assured Destruction through nuclear conflict and the Cuban Missile Crisis of October 1962. Game theory is often considered part of economics, with 11 economists being awarded Nobel Prizes for their work in the area. However, it has also been used to understand biological problems such as evolution and competition between species.

In keeping with this part of the book, it's worth thinking that while they compete directly over resources, players also compete over information. In many game-theoretic settings, having information about your opponent's strategy can help you design a strategy to beat them (just as in Rock Paper Scissors). As a result, often players will aim to design a strategy which, even if known, cannot be beaten, just like the independent and uniform Rock Paper Scissors strategy.

Prisoner's dilemma

The classic game theory problem is the prisoner's dilemma, which is usually phrased as follows, though the exact numbers vary. Two prisoners have been arrested for the same crime and are being held in separate rooms in a police station. There isn't sufficient evidence to convict them of the main crime, but there is enough to convict them of a less serious crime. The police offer them a choice: either they can implicate the other person in the main crime, or they can choose to remain silent.

If they both remain silent, they can only be convicted of the less serious crime and will both serve one year in prison. If they both betray the other, they will both serve three years in prison. If one stays silent and the other chooses to betray, then the silent partner will serve five years in prison as the sole person responsible and the betrayer will walk free.

If Prisoner A is a perfect logician, they will reason as follows. There are two possible situations:

1. Prisoner B has chosen to betray me. If I stay silent then I will serve five years in prison, if I betray them then I will only serve three. Therefore, it is better for me to choose to betray.

2. Prisoner B has chosen to remain silent. If I stay silent then I will serve one year in prison, if I betray them then I will walk free. Again, it is better for me to choose to betray.

Looked at purely through self-interest, whatever Prisoner B does, Prisoner A's outcomes are improved by deciding to betray. However, if Prisoner B reasons the same way, then they will come to the same conclusion since the outcomes are the same from their point of view. Hence, by rational argument, Prisoner A and B can talk themselves into a situation where they both betray and both serve three years in prison, whereas if they had both stayed silent their sentence would have been much shorter. Through logic and being unable to trust the other, they both ended up with a worse outcome. The key is that the prisoners being held separately means that they are unable to share information and cooperate.

Rather like Rock Paper Scissors, this prisoner's dilemma scenario has been studied intensively, both from a theoretical and practical standpoint. In fact, just like Rock Paper Scissors there are tournaments where computers play repeated prisoner's dilemma scenarios, allowing the programmers to experiment with new strategies. In practice, people don't always reason quite so coldly, and are often prepared to remain silent. There is a version where the scenario is repeated across multiple rounds, in theory allowing players to build trust in each other and achieve the better outcome together as the repeated games continue.

One particularly simple strategy that performs well is referred to as Tit for Tat: a player stays silent on the first round, and after that simply copies their opponent's previous move. Clearly two Tit for Tat players will stay silent forever, both doing well. However, someone who is not playing this strategy and who chooses to betray will essentially be punished for it in later rounds.

A version of the prisoner's dilemma was observed during the Cold War, as the United States and Russia both increased their defence spending. We can think of doing this as corresponding to the 'betray' option and leaving defence spending where it is as 'stay silent'. If both sides increase their spending, they will both end up in a worse situation than if they had followed the status quo, whereas one side spending more can dominate an opponent who does not.

For this reason, it appeared rational for both sides to keep ratcheting up their spending, and the situation could only be defused by a series of talks where either side could learn to trust the other. For example, disarmament was possible by each side showing that they could keep their promises to disarm by a certain amount, with future rounds of talks having access to the information that the promise had been kept, just like the Tit for Tat strategy above.

We can think of questions related to pandemics in a similar way – for example, the question of how strictly an individual should follow advice to stay at home. From an individual point of view, it can appear like a rational decision to go outside, since if everybody else stays at home then the risk to an individual is extremely low. However, of course if everyone reasons in the same way, then everywhere will be crowded, and the risk will become much higher. Such phenomena are sometimes referred to as the Tragedy of the Commons, referring to the competition for shared grazing land, where once again it is rational for everyone to behave in a selfish way which collectively has unpleasant consequences.

Zero-sum games

The prisoner's dilemma is an example of what we refer to as a game. It has two players who each have clearly defined choices, according to which they both receive a pay-off. Knowing the pair of choices that the two players make we can determine the reward or punishment they both receive.

The prisoner's dilemma deals with issues of cooperation and collaboration, but there is another widely studied class of games where cooperation is not possible. These are referred to as *zero-sum games* and have the property that for any pair of choices one person's pay-off is minus the other person's pay-off. We can think

Game theory

about this in terms of money – Player A needs to pay or receive a certain amount of money from Player B. Think, for example, of a game of poker, where the players' wallets always contain the same total amount of money, but where sums can be moved from one to another between rounds. Since Player B's pay-off is always minus Player A's, we don't need to think about both their perspectives and can describe the pay-offs in terms of what Player A would receive in each circumstance.

Consider the following game. Player A and Player B both sit in front of a machine, and each has a lever to move – Player A's lever can point at positions marked Left and Right, Player B's lever can point at positions marked Up and Down. The position at which they each place their levers determines how much money Player A receives from Player B. For example, if Player A picks Left and Player B picks Up, then B will pay £5 to A. If Player A picks Right and Player B picks Down, then B will pay £7 to A. We can summarise all the outcomes in a table:

	Player A picks Left	Player A picks Right
Player B picks Up	£5	£4
Player B picks Down	£3	£7

Obviously, Player A wants to maximise the amount of money they receive from Player B, and Player B wants to minimise the amount of money they pay out. Now, we can think about the right strategy, in a similar fashion to the prisoner's dilemma game above.

Suppose Player A always picks Left. Then, Player B should

273

always pick Down, to make the money paid out as small as possible. But if Player B always picks Down, then actually Player A should pick Right. But if Player A picks Right, then Player B should pick Up. It feels as if the situation is jumping around all over the place – nobody can agree on the right strategy.

In fact, John von Neumann had the fundamental insight that we shouldn't think about a strategy as a single choice of Left or Right, Up or Down. In fact, both players should play at random. Before making their choice, Player A should toss some biased coin, and if the coin comes up Heads they should pick Left, and if it comes up Tails they should pick Right. Similarly, Player B should toss a coin with a different bias, to determine whether to go with Up or Down. The technical terminology of always picking either Left or Right is a *pure strategy*, whereas choosing randomly of called a *mixed strategy*.

This is a remarkable insight, and extremely surprising. One might imagine that one choice should be fundamentally better than the other, and there should never be any disadvantage in taking that option. Von Neumann realised that the best strategy is not to be too dogmatic, and to keep one's options open. If we play this game very many times, by the Law of Large Numbers we know that Player A will pick Left a certain proportion of the time. In general, this proportion need neither be zero nor one. What this illustrates is the value of balance, or not going in too hard with one choice in this case.

The question is how biased the coins should be? I will tell you the answer in this case and explain why it is an attractive one. The key is to give one's opponent no incentive to switch their strategy. As in Chapter 5, we work in terms of expected value.

Game theory

Player A should use a biased coin that leads them to pick Left 3/5 of the time and Right 2/5 of the time. To see why this is a good strategy, consider Player B's two choices. If Player B picks Up, then the expected value of A's winnings is 3/5 x 5 + 2/5 x 4 = £4.60. If Player B picks Down, then the expected value of A's winnings is 3/5 x 3 + 2/5 x 7 = £4.60. In both cases this is the same pay-off. In other words, whatever Player B does, A can guarantee that on average they will win £4.60, which is attractive because there is no larger amount that A can guarantee themselves.

Similarly, Player B's optimal strategy is to pick Up 4/5 of the time and Down 1/5 of the time. Again, this is best because the two expected pay-offs match, being 4/5 x 5 +1/5 x 3 = £4.60 and 4/5 x 4 + 1/5 x 7 = £4.60 when A picks Left and Right respectively. By making this choice, B has limited their losses – there is no smaller amount that B can arrange to have to pay out. Indeed, this property of ensuring that the pay-offs match is how we know that these numbers, 3/5 Left for A and 4/5 Up for B, are the right ones. We look for a mixed strategy where nothing that the other player does will affect the pay-off.

This is an example of a Nash equilibrium, named after the mathematician John Forbes Nash, who was the subject of the book and film *A Beautiful Mind*. In such a situation, all players have no incentive to change strategy, and the game is in a stable state compared with the 'jumping around' situation I described above for pure strategies. Nash showed that such a situation arises much more generally, for a wide variety of games with more players and more options.

Minimax strategies

In fact, we can understand Player A's optimal choice of coins by plotting the following graph, which represents the situation from their point of view.

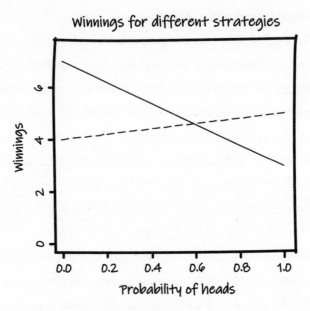

Winnings for different strategies

I have plotted probabilities from 0 to 1 on the x-axis. This represents a range of possible biased coins that A could use, ranging from one that always produces Tails (on the left-hand end) to one that always produces Heads (on the right-hand end). A fair coin lies exactly in the middle. There are two lines plotted, representing the expected value of A's winnings according to what move Player B makes.

1. If Player B chooses Up, then A's expected winnings with different coins will be given by the dashed line, running from 4 on the left-hand side (corresponding to using a coin which always comes up Tails, so A must pick Right) to 5 on the

right-hand side (corresponding to a coin which always produces Heads so A must pick Left).

2. If Player B chooses Down then A's expected winnings will be
given by the solid line, running from 7 on the left-hand side
to 3 on the right-hand side.

We can imagine that Player A has picked their coin in advance
and that this information is known. Player B can take advantage
of this strategy. If A picks a coin towards the left-hand end of the
range, Player B can minimise the amount they have to pay out by
ensuring that the winnings are on the dashed line – so in other
words Player B should pick Up. If A picks a coin towards the right-
hand end, Player B can minimise the amount they have to pay out
by forcing the winnings to be on the solid line – so Player B should
pick Down.

Overall, if Player B is sensible like this, then A's expected pay-
off will be given by the upward pointing triangle formed by taking
the dashed line up, and then the solid line down. This triangle is
highest precisely at the point where the two lines intersect, which
corresponds to the coin with a 3/5 probability of coming up Heads.

This is often referred to as a *minimax strategy*. That is, we look
to minimise the maximum of something. In fact, from Player A's
point of view this is really a maximin strategy, where we seek to
maximise the minimum pay-off. That is, whatever A's choice of
coin, Player B will choose their strategy to take the minimum of
the two lines. So, A should look to make this minimum as large
as possible. In this way, they negate the value of Player B knowing
which coin they will use and stop them exploiting the potential
value of this information.

Numbercrunch

We could draw a similar picture from Player B's point of view. The only difference would be that B is looking to minimise the pay-off. However, the same kind of picture would show that B's best choice is where the two corresponding lines intersect, which corresponds to the coin with a 4/5 probability of Heads. In fact, I encourage you to draw the picture and check this.

One more nice feature of this picture is that it makes it clear why the right strategy lies somewhere in the middle. If A picks Left then the pay-off is greater for Up than for Down, whereas if A picks Right then the opposite is true. Whenever this is true, then the picture will contain straight lines which cross at some point in the middle, and so a mixed strategy will be best. The only issue is to determine where this intersection takes place.

We can think about dating apps such as Tinder using ideas related to game theory. Suppose that each individual can swipe right or left on a user's profile to accept or reject them as a potential partner, and a match will occur only if both people swipe right. With the crude aim of maximising the number of matches, it makes sense to swipe right every time, to avoid missing any potential match. However, clearly if everyone adopted this strategy then every potential pairing would lead to a match, and chaos would occur. It therefore makes sense to be more selective – in fact, data suggests that female users in particular tend to do this, thus making the system workable.

But how selective should you be? Each dater may informally score each profile based on the characteristics most important to them, such as appearance or shared interests, for example. A natural strategy is to have some quality threshold, and to simply swipe right on each profile whose score exceeds that threshold. However, it may not be obvious how to choose this threshold. In fact, mathematical ideas related to the so-called Secretary Problem[11] might suggest that the right strategy is to swipe left on a fixed number of profiles to start off with, to calibrate your expectations of the partners that might be available and to determine a sensible threshold. Of course, other non-mathematical approaches may be possible!

Should I stay or should I go?

Based on similar ideas, we can think about a simplistic toy mental model that might help us think about everyday problems, and which suggests that a balanced approach to life may be valuable. Note that, strictly speaking, this is not a game like the prisoner's dilemma in the sense that there is really only one player – you can think that the other player ('Nature' or 'the state of the world') has fixed their strategy in advance.

Imagine that you have been invited to a party, which runs from 9pm until 3am the next day. It will be a fantastic event with wonderful food and drink, a chance to celebrate with your friends and to have a good time. Unfortunately, the next day you must also give a presentation to your boss at 9am, which will help determine your career prospects. The question is what time to leave the party.

We can represent the time of departure on the x-axis of a graph, and the harm resulting from your decision on the y-axis. If you

---

11 In this setting, we simply wish to hire the best person for a job. We will see a large number of candidates, for example 100, and must make a decision whether to hire each one immediately after the interview, without knowing the strength of the overall field. It turns out that the optimal strategy is to just observe the first 37% of the candidates, and then to make an offer to the first candidate after that who is better than everyone you have seen already.

solely focus on the question of fun at the party, it makes sense to stay as long as possible, because your mental map looks like this – the later you stay, the more chance you have to see your friends, the more you can eat and drink. If you don't go at all, you'll miss out on this. Maybe there's a law of diminishing returns – as time goes on, you've seen the people you really want to see, food and drink is running out, and the additional value of staying later reduces. However, purely from the point of view of minimising party FOMO, you ought to stay until 3am.

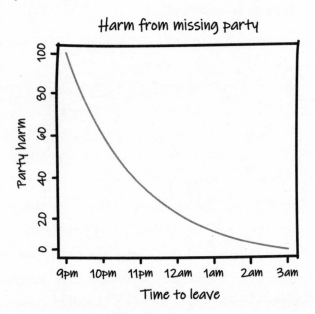

Harm from missing party

Time to leave

Whereas if you think about harm to your presentation, and hence to your career, the picture perhaps looks something like the one on the following page. It doesn't matter too much if you go at the beginning of the party, but as time goes on you will lose sleep, you are more likely to arrive at work feeling rough, and the quality of your presentation is likely to suffer more the later you wait. Purely from the point of view of work, you ought to leave at 9pm (or skip the party altogether).

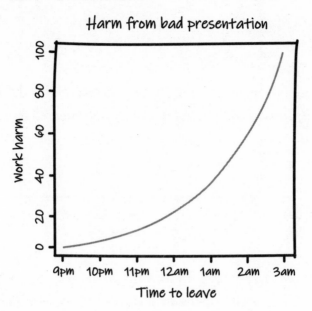

However, it seems to me that the answer is that we need to consider both types of harm and add together both types of negative consequences. Notice that this is slightly different from the example above, where we needed to take the minimum of the two straight lines rather than add them together. However, the principle is somewhat similar – adding the curves together you end up with the dark curve, and the optimal point is somewhere in the middle, representing a mixed strategy.

I haven't said what units I'm measuring harm in. The curves may well not be the same height, meaning that the optimal time to leave may not be exactly in the middle. In fact, for different people, the curves will be different, according to their stamina, how good their presentation skills are, which friends will be

Total harm

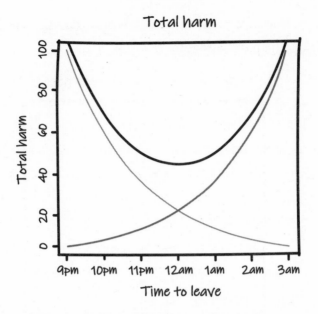

there and how forgiving their boss might be. This is certainly not meant to be a rigorous mathematical proof of the virtues of work-life balance. However, it does illustrate that ideas such as game theory can offer a mathematical way to justify a balanced approach, not just in this toy setting but in a range of other scenarios as well.

This is a simplified version of the true problem, assuming we decide our leaving time before even arriving, which is referred to as a *non-adaptive* strategy. In reality we may change our plans as the evening goes on. Instead of essentially estimating the curves in advance, we will gain more information about them over time. Perhaps the people you really wanted to see aren't able to be there. Or perhaps you hit it off with a total stranger and want to stay longer. This new information would change the curve for party harm, and so you could use an *adaptive* strategy, which updates your best move based on what you learn.

Game theory

Of course, there are constraints on the strategies available to you: you can't travel back in time. If at 11:30pm you are having a terrible time because you inadvertently insulted the host ten minutes ago, you can't suddenly change the strategy to leave at 11:15pm. However, for a wide variety of problems this distinction between non-adaptive and adaptive strategies can be crucial – the flexibility offered by being able to adapt can lead to much better outcomes. When playing Wordle, for example, your strategy is highly adaptive, taking account of letters learned from previous guesses – a non-adaptive strategy of fixing your six guesses ahead of time would be extremely unlikely to succeed. In the same way, the ability to test adaptively can make significant improvements to the performance of the pooled testing algorithms I talked about in Chapter 9.

A similar argument of balancing harms might apply to the question of how forcefully a government should intervene against an infectious disease such as COVID. If you solely focus on COVID deaths, it would have made sense to intervene as hard as possible, to minimise the number of these. Of course, there may well have been other consequences to lockdown in terms of health consequences, in terms of missed diagnoses of diseases, mental health issues, isolation for dementia patients and so on. Equally if you solely focus on the harm to the economy, the more you keep things open, the better things are. Of course, if the disease had really been allowed to run riot, then there would have been a danger of the breakdown of civil order and other economic consequences. Again, an argument based on balancing harms might suggest that a response somewhere between these two extremes might have been optimal.

Reinforcement learning

Indeed, ideas from game theory have been applied to develop the kind of modern machine learning or artificial intelligence algorithms which I have already referred to. Specifically, the field of reinforcement learning is motivated by the idea that a player of this kind of game may not even have information about the values in the table of pay-offs at first. This can represent a computer in a situation of complete ignorance about its environment.

However, by making a series of choices, the computer can start to work out what pay-offs are associated with which choices, and gradually start to gain information about the structure of the possible rewards available to it. Once it has done this, it can start to move from this initial learning phase into a state where it has a good understanding of its environment and can start to take the best actions based on the information it has gained.

These kinds of reinforcement learning algorithms were used for example by Google DeepMind's AlphaGo program, which in 2016 famously achieved what had previously been believed impossible by beating an expert human Go player, Lee Sedol. Part of the training that AlphaGo undertook essentially consisted of reinforcement learning where, following study of a database of previous historical Go games, the program experimented and practised by playing a large number of games against itself in order to work out which kinds of moves could work best.

A later version, AlphaGo Zero, even dispensed with the need for an initial database, by using pure reinforcement learning to develop winning strategies from scratch purely from a list of the rules of Go. Such achievements may seem limited, in the sense that mastering a board game is only one part of the all-round general intelligence that humans can exhibit. However, the development of AlphaGo was a huge breakthrough in the performance of computing and algorithms, and one which is likely to have major consequences in our everyday lives for years to come.

I have described how game theory can be understood in terms of information, and so it is not surprising that Claude Shannon was attracted to these kinds of problems. I already mentioned in Chapter 9 that he had a hands-on and practical approach, coupled with a sense of fun, and these facets of his character often met in the arena of games. Shannon was one of the first people to seriously think about programming a computer to play chess, and the so-called Shannon number 10^{120} (an unimaginably enormous number, written as a one with 120 zeroes after it) arose as his estimate for the number of possible chess games. He collaborated with legendary gambler and investor Ed Thorp to build a computer to predict the outcome of a roulette wheel.

Shannon invented what he called the switching game, which is played on a network, where play alternates between one player who tries to colour edges to create a path connecting two prescribed vertices together while the other player removes edges from the network. This is an example of a wider class of what are known as Maker-Breaker games, which are still an active topic of research for game theorists.

In all these ways, the topics of information, networks and game theory which I have described in Part 3 of this book are linked together as part of a rich tapestry and understanding them better can help us make sense of the modern world. However, while we do so, it is good to remember Claude Shannon's sense of fun and of exploration.

Summary

In this chapter we have seen how game theory can describe competitive interactions in a variety of settings, including nuclear stand-offs and the classic prisoner's dilemma. By thinking about information and pay-offs, particularly in the setting of zero-sum games, we were able to formulate the idea of mixed strategies. We argued from the minimax principle that such strategies negate the value of the other player having information about the strategy itself and argued that arguments about mixed strategies may suggest the value of a balanced response to real-world problems.

Suggestions

You might like to explore the ideas of this chapter further, for example by looking for everyday situations that can be formulated in this setting. If you have two children, can you create a (positive!) version of the prisoners' dilemma, where they will both be rewarded with two sweets if they both tidy their rooms, whereas if only one tidies then that child alone will get one sweet? What effect might that have? You might also like to think about mixed strategies more by populating the reward table of the zero-sum game with different numbers. Are there examples where a pure strategy always wins? (You might like to think about what happens if the solid and dashed lines do not intersect, for example.)

PART 4:
LESSONS

Chapter 13
Learning from error

I have described a collection of mathematical ideas which are useful to make sense of the world. I divided these into the categories of Structure, Randomness and Information, although clearly there are overlaps and interactions between these themes. Many of these ways of thinking that I have described proved useful in particular ways during the COVID pandemic, but the next big news story probably won't require exactly the same tools to make sense of. It is quite possible that exponential growth and random walks may not be as big a factor, for example, though I passionately believe that a strong grasp of Structure, Randomness and Information will prove invaluable on many occasions in the future.

On the other hand, there are more general principles that we can abstract, having watched people develop both formal mathematical models and informal beliefs that have not always survived contact with the real world.

Numbercrunch

1. Consider your assumptions

The first suggestion is that you should always question your assumptions and the predictions arising from them. For example, as I have described previously, many people convinced themselves that the rise in European coronavirus cases in the summer of 2020 without a corresponding rise in deaths meant that the virus had become less dangerous. This could have been avoided by a more careful look at the data, examining the age range of cases and perhaps plotting trends on log scales.

But in general, whatever you believe about the world is likely to be based on a series of logical deductions, 'If X then Y, if Y then Z'. Mathematicians are used to thinking in this way when developing proofs of theorems out of a chain of smaller results. Every researcher often experiences the sinking feeling when one link in the chain turns out not to be correct, meaning that the whole series of inferences falls apart. It is good practice to try to think in the same way – just as with Fermi estimation, to break down your beliefs into a series of smaller parts. If you can think about how robust each of these parts are, and how further evidence that invalidated one of them would affect the whole thing, then you will base your decisions on more than just gut instinct. What is the weakest part of your argument? What is the strongest argument someone could make against it? If you can't answer either of these questions, there's a good chance that you are fooling yourself as to the rigorousness of your case.

Nobody likes to admit that they are biased, but ultimately none of us were dispassionate when considering issues as complicated as the coronavirus pandemic. We all have personal circumstances (different levels of risk for our family, a range of economic vulnerability depending on our employment status, and varying degrees of mental and physical health) affecting our judgement. It is inevitable that some of these things filtered into our assessment of the health risks of the virus against the economic risks of measures taken to combat it. But at least by understanding your own circumstances, you can question to what extent you are representative of the wider population and how to balance the varying needs of the entire community.

2. The world is a messy place

When comparing coronavirus data between countries, many people wanted to find a 'magic bullet' – a single factor that explained everything. This is almost certainly not a realistic aspiration.

It was possible to think of lots of factors that might affect the spread, even before we started to consider a government's response. These factors may include the age profile of the population, the population density (perhaps in a 'lived' form that considers the experience of a randomly chosen person), demographics of typical households, underlying health conditions, climate, pre-existing levels of exposure to similar viruses, degree of geographical isolation, typical levels of international travel, cultural factors surrounding willingness to wear masks, individualism (how likely the population is to obey government instructions), level of laboratory infrastructure, timing of the epidemic relative to other countries, and simple blind luck.

This already bewildering list of factors is far from exhaustive, and so you should be distrustful of naive and simplistic analyses which claimed that 'countries with female leaders handled the pandemic better', without attempting to take account of other effects. As we have seen in Chapter 6, data is itself noisy, and as in Chapter 11, the way in which different countries measure and report it has an enormous impact on outcomes.

Even when we consider a single country or region, things are always more complicated than you might hope, and measures always have second and third order consequences. For example, there may be data suggesting that pubs were a major source of infection, and so should be closed. However, this may reflect the way that data is collected (people remember being in the pub but may not remember the person in the supermarket queue who infected them), and pub closures may have led people to drink in other people's houses instead or to adopt other riskier behaviours as a result.

All these issues mean that attempting to predict or explain such a complicated story such as the pandemic was a bewildering task, and the same is likely to prove true for future crises. This is not to say that we should not try to do so. However, it is worth remembering that predictive models come with uncertainty, and that any analysis that attempts to explain the whole situation using a few factors is almost certainly too simplistic.

3. Don't overweight the past

When considering an issue, it is natural to look for examples of how apparently related stories have played out, to judge how it might happen this time. However, remember from Chapter 1 the danger of overfitting, where we tried to join the dots through a single curve. Such a model is likely to be wrong because it takes too much account of individual data points which are noisy and uncertain.

Learning from error

There is a considerable danger of recency bias (fighting the last war) and of over-correcting for past mistakes. An excellent example of this came through many pundits' predictions of the 2020 US presidential election. Many of these pundits had been very wrong about the 2016 election, giving Donald Trump no chance of winning and assuming that opinion polls were completely accurate. When they came to predict the 2020 election, it was natural to take this into account. However, no situation repeats itself exactly, and many people over-corrected their views in the opposite direction, by now assuming that polls were worthless, and that Trump had a roughly even chance of winning. This was reflected in the betting markets, which allowed punters to bet on Trump at odds of 2/1 or so (having a 33% chance of winning as described in Chapter 8), at times when statistical analysis of the polls showed that the chances were much lower.

In fact, despite a degree of polling error and delays in reporting the results of certain key states, Biden won comfortably by both popular vote and electoral college. A more level-headed analysis, taking account of the fact that polls and statistical analysis had worked very well in 2008 and 2012, would have not taken the lessons of 2016 so seriously to heart.

In the same way, many people downplayed the danger of the coronavirus since diseases such as MERS, SARS, Ebola and swine flu had dominated the headlines for a few weeks without causing such major worldwide disruption. Again, it was reasonable to take these recent experiences into account, but prudent not to overfit to them, and consider whether things could be different this time.

Equally, when considering the European and North American

monkeypox outbreaks which started in May 2022, it would be prudent to not assume that this would be a repeat of COVID, in terms of seriousness, method of spread or political response. There are lessons to be learned from the coronavirus pandemic, but the same scenarios are unlikely to repeat exactly, and it's sensible to consider ways in which monkeypox and COVID differ.

4. Don't cherry-pick

When examining the evidence regarding COVID or other issues, there is a wealth of scientific data available, much of it published in peer-reviewed papers. In an ideal world, this would be the gold standard: papers would only be published that had passed a quality threshold and that were based on a rigorous analysis of publicly available data. Sadly, this is not always the case.

The world of scientific publishing is unfortunately not all that we might hope. There are very many so-called predatory journals, which appear to an outsider to be legitimate, but which will essentially publish anything whose author is prepared to pay a certain fee. If this seems like an exaggeration, there have been several incidents where nonsense deliberately generated by computer has been accepted for publication by journals of this kind.

Unfortunately, even within legitimate journals, passing peer review cannot be seen as a cast-iron guarantee that the results are correct. Reviewers are generally busy academics, who are not paid for their role, and often fit in refereeing papers around their own research and teaching duties. Reviewers and editors generally do their best in difficult circumstances, but they cannot be assumed to be gatekeepers of absolute truth, particularly given the possibility of authors acting in bad faith to obfuscate the truth. In general, a paper having passed peer review is a good sign, but some such papers later being retracted indicates that it is not an infallible process.

However, even if we only consider peer-reviewed results, we will still find a problem. Take, for example, the Infection Fatality Rate (IFR) for coronavirus. There are many papers that seek to estimate this, but which all find different answers, which is to be expected for several reasons. Firstly, as I have previously described, the IFR depends on the age of the population and the standard of healthcare, so surveys carried out in different places are estimating different quantities. Secondly, since we can never know how many people were infected with coronavirus, the result of any such study will require estimating an unknown quantity, which also introduces potential error.

In general, differing IFR estimates are synthesised via what is known as a meta-analysis, where these separate papers are considered together and combined into a single estimate. This is not carried out by crude averaging, but by a more sophisticated process involving rating each individual paper separately (for example, by the size of the sample) and giving them a different weight in the overall calculation.

Even this meta-analysis process is not perfect, because the process of rating and weighting is clearly somewhat subjective. However, it is better than a common alternative, which is to simply choose the study you like and use that result. This is informally referred to as 'cherry-picking'.

Indeed, we can safely assume that given enough surveys, it is extremely unlikely that the most extreme values in either direction will be accurate. Just as the highest and lowest marks from figure skating judges should be removed, we should look for the truth somewhere in the middle. In that sense, if someone quotes a value for IFR from a published paper in a journal with a prestigious name, that may not be enough: it is necessary to consider the wider context of the literature.

We can think of this in terms of our coin experiment. If we performed 50 experiments, each consisting of tossing a fair coin 100 times, the table in Chapter 5 shows us that we would not be surprised to see as few as 40 or as many as 60 Heads showing some of the time, just by random fluctuations. However, to estimate the fairness of the coin by simply cherry-picking the experiment with the lowest or highest number of Heads would clearly be dishonest and misleading.

5. Models can only go so far

A quote often attributed to statistician George Box is: 'Statisticians, like artists, have the bad habit of falling in love with their models.' What he perhaps meant by this is that having invested time in developing an explanation of the world, we can remain convinced that it is right even as the evidence builds against it. This can apply both to formal mathematical models developed from data, or to a more informal understanding of the world based on reading various sources. Once we believe that a model is right, it can take extraordinary evidence to overturn it in our mind, which can become more of a matter of faith than of justified belief.

We should remember the other quote attributed to Box that I

mentioned in the introduction: 'All models are wrong, but some are useful.' What he meant by this is that, as we have seen, the world is a complicated place, and it seems very unlikely that any simple set of equations involving a few terms and parameters can summarise it perfectly. However, nonetheless, simple models can often work remarkably well for a long time.

For example, the model of unchecked exponential growth in North West hospital patients I described in Chapter 6 gave predictions that were valid for six weeks or more, long enough for it to be clear that there was a problem. Clearly this kind of exponential growth cannot go on forever since there are only so many people and so many beds. Hence, the trick is to appreciate that while a straight line fits the data for now, we cannot extend it indefinitely. Obviously, this is a vague, and almost platitudinous, statement, but it indicates the value of thinking about the limitations of a model: if it works well in the North West of England in autumn, it seems reasonable to think it might work in the North East at the same time of year, but do we believe it would be valid in an Australian summer?

However, returning to Box's theme of falling in love with a model, it is worth remarking that something being complicated or demanding to create does not make it any more likely to be true. Often, even subconsciously, you might feel that the time you have personally invested in computations, or a data visualisation, must be worth something, so there must be value to it.

Coupled with the point about cherry-picking above, this leads to the danger of confirmation bias. We are generally much happier to accept facts that confirm our prior beliefs than facts that contradict them. Clearly, there is value to attempting to maintain a consistent world view, but equally you should be open-minded to the possibility that the data has started to move away from the previous trend, and that your ideas may need to be updated.

6. Consider the possibility of groupthink

When communities work and discuss issues together, it is natural that they will reach a shared understanding of the problem and a group solution that they all feel comfortable with. This is generally a good thing, which means that they have considered the issues and come up with a sensible solution. However, there are certain dangers too, from the point of view of groupthink.

There can be social pressures or an assumption that the problem is solved, which mean that the solution, once established, is never questioned. Further, the next time an apparently similar issue arises, it can be natural to assume that it is identical to what has previously been discussed, and to simply adapt the old solution to the new problem rather than approaching it with fresh eyes.

These are obviously dangerous ways of thinking, which can lead to complacency. It is already hard for one individual to question their assumptions, as I have argued above that they should, but the collective mindset can be harder still to change, particularly in a large or traditionally minded organisation. There is certainly a role for people to play devil's advocate, and indeed such interactions should be encouraged.

Indeed, the problem of groupthink can be seen as an issue for the wider scientific community. Very often new ways of thinking and rejection of old ideas are difficult to achieve within a particular field. This effect is accentuated by the way that scientific communities evolve: the people who had the old ideas are rewarded for them by promotion to senior positions, editorships of journals and service on grant-awarding panels. Such people are, by definition, invested in the status quo and may not remain open-minded regarding questions that were apparently already settled.

There is a delicate balance here, of course. I am not arguing for iconoclasm that tears down the established scientific consensus every time someone feels like a change. Indeed, much of the discussion around COVID has shown that many established ideas, including the simple SIR models I described in Chapter 4, have lasted the challenges of modelling better than many novel theories created to overturn them. The odds are very much in favour of established theory, despite how energetically contrarians argue against it. As a result, it would be prudent for people proposing a new theory to engage respectfully and thoughtfully with the existing literature, and for scientists moving into a new field from outside to listen carefully to the wisdom of those trained in it.

However, it is vital for the health of a community that novel ideas can be heard and tested, however unlikely they may sound. As a sensible way to proceed, we might assume that extraordinary claims require extraordinary evidence. That is, even if a theory does not need overturning, we may see random fluctuations that suggest problems with it. However, we cannot simply dismiss innovative ideas just because they sound stupid: it is probably highly likely that many of the current established ideas once sounded stupid when compared to the consensus of the time.

7. Wishing doesn't make it so

At various times in the coronavirus pandemic, it was probably reasonable to have differing levels of optimism regarding the prognosis of the situation. It was a challenging time for many people in terms of the personal and emotional struggles that the disease imposed on everyday life, and it was natural to want it all to be over.

However, wishing that the pandemic would go away did not make it happen. In the same way, if you stare at a graph for long enough to convince yourself that the numbers are going down, it does not make any difference to the future points. As the great physicist Richard Feynman wrote in his report on the Space Shuttle Challenger disaster: 'Nature cannot be fooled.'

There were various times when considering competing coronavirus models and theories, the right answer was simply that 'we will know soon enough'. It is possible to devise any number of apparently elegant theories that do not survive prolonged contact with the real world, and so elegance of the theory alone is not sufficient grounds to decide its truth.

However, a reasonable question to ask is: 'What predictions does this theory make?' A proper scientific theory will give answers that can be tested against future data, and if the data turns out to not be consistent with that, then the theory cannot be true. Perhaps it can be adapted, but if we predict Herd Immunity at 20% and see infections reaching 30% of the population, then it may be that we must hold up our hands and admit defeat.

Learning from error

8. Humility - admit your mistakes

It's very likely that at some stage you got something wrong, possibly in public. However, there is no shame in this. Without exception, nobody was right about every aspect of the coronavirus pandemic. Given a new situation of this kind, with a virus spreading rapidly across the globe through some combination of airborne and contact transmission, including the possibility of asymptomatic spread, with an unprecedented solution of lockdowns being imposed to varying degrees by governments, and with technology such as PCR tests and novel forms of vaccine being used to fight it, it was not reasonable to think that anyone could call everything correctly first time.

Indeed, reducing a complex situation down to a simple sound-bite or two is probably a good sign of suspicion when evaluating a scientist or pundit. As I have described, the world is messy, and so is real data, and it seems unlikely that a complicated problem has a simple solution, or that one person could have a 100% success rate on every analysis of every aspect of it.

Being wrong is not a mark of shame in this respect. A few wrong calls on one front do not invalidate valuable contributions made in other areas by the same person. However, the way that someone responds to their mistakes is important. To continue to double down on the wrong analysis, or to obfuscate and deny having been wrong in the first place, is an extremely unhelpful (if natural) reaction.

It is better to simply admit error, perhaps reflect on why this occurred, and to move on. While there appears to be a stigma attached to politicians performing a U-turn, it is surely better for scientists to make that turn and to start to move in the right direction.

9. Centrism and moderation

Throughout the coronavirus pandemic, I tried to advocate a centrist approach. As I have described above, it is unlikely that the extreme estimates (in either direction) of a quantity are the most accurate ones. In the same way, in a particular situation, given a range of analyses ranging from 'end of the world' to 'nothing to worry about', it is likely that the right answer lies somewhere in between.

Similarly given a range of solutions from 'lock down as hard as possible for as long as possible' to 'do nothing', we might be suspicious of the effectiveness of answers from one extreme or the other. As I have described, this is a complex situation, and any proposed answer has drawbacks and trade-offs associated with it. By choosing an extreme position, these can be made worse.

This does not imply that we should blindly pick a position halfway between the most extreme opinions. In many cases, being extreme in one direction or another can be the right response. However, it's certainly worth reflecting on whether this is true and trying to understand the other extreme without demonising or caricaturing their position, before deciding to take this kind of response.

Similarly, in communicating data, it is important to emphasise the nuances and uncertainty around it. We cannot know everything for sure, but it is probably unlikely that things are as bad or as good as they first appear from an apparently outlying sample of data. In other words, the principle of moderation until further data appears can prevent over-reaction in one direction or another.

Indeed, there is an argument for centrism and moderation in estimation as well. The lesson of Fermi estimation from Chapter 2 is that we can obtain a sensible answer to a question by combining a collection of imperfect estimates. However, the Fermi method assumes that we combine the values in the middle of the plausible range, rather than extreme values. If we combine each of the most extreme estimates in one direction or another, the final answer is likely to represent a worst- or best-case scenario. While it can be useful to know what this is, we shouldn't let such scenarios dominate our thinking or come to believe that they are the most likely outcome.

10. Maths is the right tool

My final message is that mathematics is likely to be the right tool to make sense of the situation. Whether in understanding how functions grow, the role that randomness and uncertainty play, or what Information Theory tells us about filter bubbles and correlated information, mathematical techniques can give insights in a way that is divorced from emotion and personal biases. The key ideas of Structure, Randomness and Information provide powerful tools to add to your thinking.

While some of these tools are not studied until the later years of a university mathematics degree, it is not necessary to have such training to understand the principles that I have described. By developing a sense that it is important to ask, 'are these numbers reasonable?' or 'what is the margin of error on that figure?', anyone can use mathematical principles to think about the world in a smarter way.

I believe this is not a coincidence. The coronavirus itself did not know advanced mathematics, but it spread according to an internal logic that can be studied with the same mathematical tools that allow us to study properties of objects in space, or dice rolls, or random walkers staggering home from the pub. In the words of Richard Hamming, in his essay 'The Unreasonable Effectiveness of Mathematics':

'During my thirty years of practicing mathematics in industry, I often worried about the predictions I made. From the mathematics I did in my office I confidently (at least to others) predicted some future events – if you do so and so, you will see such and such – and it usually turned out that I was right. How could the phenomena know what I had predicted (based on human-made mathematics) so that it could support my predictions? It is ridiculous to think that is the way that things go. No, it is that mathematics provides, somehow, a reliable model for much of what happens in the universe. And since I am able to do only comparatively simple mathematics, how can it be that simple mathematics suffices to predict so much?'

Acknowledgements

I would like to thank everyone who encouraged and supported me in the writing of this book. I received thoughtful and constructive comments on various drafts from Christine and Roger Johnson, Paul Mainwood, Annela Seddon, James Ward and Sarah Young. Matt Aldridge, Steve Forden, Simon Johnson and David Leslie patiently helped me make sense of various issues.

I'm lucky that my employer, the University of Bristol, believes firmly in public engagement and have received particular support from Jonathan Robbins, Jens Marklof, Liz Clark, Chrystal Cherniwchan, Victoria Tagg and Philippa Walker. Stuart Ritchie, David Sumpter, Remi Lodh, Diana Gillooly and Hallie Rubenhold provided invaluable advice on the world of commercial publishing.

My enthusiasm for writing was rekindled during the pandemic. I'd like to thank the sensible voices of centrist COVID Twitter, including the people who believed in plotting the data first and drawing conclusions afterwards and all the '-ologists' who didn't mind answering my stupid questions in DMs - you know who you are. Thanks also to everyone who decided the pandemic was a good time to read more about maths, or who just liked the bunny. I'm also grateful to Fraser Nelson and Evan Davis, who first gave me a bigger platform to talk about the numbers, and to all the journalists who connected me to a wider audience.

It has been an immense pleasure to work with everyone at Heligo and Bonnier Books in producing this final published edition. In particular, Rik Ubhi has shown boundless enthusiasm for the project and countless imaginative ideas to improve my drafts. The finished product looks as good as it does thanks to the cover design by Nick Stearn, and the illustrations and typesetting by Graeme Andrew. I'd also like to thank Justine Taylor, Abi Walton, Frankie Eades and Ian Greensill for all their help.

My agent, Will Francis, saw that a (very different) early version of this book had potential that I had missed, and helped to guide and shape it into the finished form you see today. I'd like to thank him for all his advice and experience, and the rest of the team at Janklow and Nesbit for taking care of so much.

Finally, and most of all, I'd like to thank Maria, Emily and Becca for their constant support.

Glossary

adaptive strategy: one that changes over time to take account of new information.

Bayes' Theorem: way to deduce probability of B given A from probability of A given B.

bit: (short for Binary digIT), a quantity that could be 0 or 1.

capacity: how much information can be sent through a particular noisy communication channel.

Case Fatality Rate (CFR): proportion of people with a positive test that go on to die of a disease.

Central Limit Theorem: repeating independent experiments enough times, the probabilies that something happens will look more and more Gaussian (bell-shaped).

conditional probability: how likely something is to happen, given that something else has happened.

confidence interval: range of possible values that we reasonably believe might be true.

data compression: efficiently representing random objects as a sequence of 0s and 1s.

degree of a vertex: number of edges coming in and out of it, in an undirected graph at least.

denominator: bottom of a fraction.

diameter of a network: number of steps to guarantee we can get from any vertex to any other.

differential equation: an expression for acceleration or speed in terms of position.

differentiation: finding the speed from a knowledge of the position.

directed: an edge of a network that can only be followed in one direction.

doubling time: how long it takes an exponential function to double in size.

edge: a line joining two vertices in a network.

entropy: measurement of uncertainty of a random quantity.

expected value (or sometimes *expectation* or *mean*): the average of the possible outcomes of an experiment, weighted according to the probability that each one occurs with.

false negative: someone who tests negative when infected.

false positive: someone who tests positive when not infected.

Fermi estimation: breaking a complex estimation problem down into small stages.

Glossary

function: a rule that we can think of as working like a machine or a computer program.

- *constant function*: one that always gives the same values
  exponential function: one that has the same amount multiplied each time
- *linear function*: one that has the same amount added each time
- *polynomial function*: one that involves square, cube, fourth power of another quantity
- *quadratic function*: one that involves the square of another quantity

Gaussian: bell-shaped, or normal, curve.

graph: either (a) a two-dimensional plot of points that we hope to put a best-fit line through or (b) a collection of vertices joined by edges. I will use graph in the sense of (a), and the word network when I want to talk about (b).

Herd Immunity Threshold (HIT): proportion of population that would need to be infected to make the R number become less than 1.

independent: events whose outcomes do not affect one another.

Infection Fatality Rate (IFR): the percentage of infected people that will die of a disease.

integration: finding the position from a knowledge of the speed.

lag: delay, both in reporting (deaths that take place today will not be announced today) and in the development of the disease (a patient infected today may not die until 28 days later).

Numbercrunch

Law of Large Numbers: repeating independent experiments enough times, the sample average will get close to its expected value.

linear regression: drawing a best-fit line through a two-dimensional plot of data to explain relationships.

log scale: graph with labels on y-axis compressed. The right way to represent exponentials.

loss: price you pay for getting a guess wrong.

Markov chain: process which has no memory, other than through its current position.

median: the middle value on a list of data ordered by size.

mixed strategy: one where a player picks their next move at random according to some biased coin.

network: a collection of vertices joined by edges.

non-adaptive strategy: one which is fixed in advance.

normal: bell-shaped, or Gaussian, curve.

null hypothesis: default assumption about the state of the world, can be overturned by data.

numerator: top of a fraction.

odds: the multiple of a gambler's stake returned for a successful bet.

ONS: Office for National Statistics – UK body responsible for compiling and releasing official data.

overfitting: trying to explain every piece of data perfectly with an overly complicated model.

Glossary

parameter: value that appears in an equation, which can be tweaked to change its behaviour.

phase plot: plot that compares position and velocity for a system governed by a differential equation.

point estimate: best guess of a particular quantity, given the data.

prevalence: percentage of people who have a disease in a particular region.

probability: how likely something is to happen.

pure strategy: one where a player always makes the same move.

p value: the probability of seeing a result as extreme or more extreme than the data, assuming our null hypothesis is true.

random: any process generated by a mechanism involving chance, or sometimes informally used for processes that seem so complicated we can't expect to model them perfectly.

random walk (aka drunkard's walk): taking steps at random but watching the position evolve.

R number: how many people each infected individual themselves infects. Greater than 1 means epidemic is growing, less than 1 means it is shrinking.

sample average: result of adding up the numbers observed from a repeated experiment and dividing them by the number of times the experiment is performed.

sensitivity: proportion of infected people who test positive.

service time: how long a customer takes to be served in a queueing model.

sigmoid: S-shaped curve.

simple harmonic motion: for example, a pendulum where force is proportional to position.

specificity: proportion of uninfected people who test negative.

statistically significant: a result which is unlikely enough (assuming the null hypothesis is true) to provide robust evidence that the hypothesis is in fact false.

time series: a list of data corresponding to a sequence of instants in time.

undirected: an edge of a network that can be followed in either direction.

uniformly random: each outcome is equally likely.

variance: how spread out the possible values of an experiment are.

vertex: a point in a network.

zero-sum game: one where player A's pay-off is minus player B's pay-off.

Further Reading

We are lucky to live in a time when there are so many great resources and people that bring maths to a more general audience.

Some websites that I particularly recommend are *Plus Magazine* at https://plus.maths.org ("bringing mathematics to life"), *Significance* at www.significancemagazine.com (for statistics and data science), and *Quanta* at www.quantamagazine.org (for accessible updates from the cutting edge of research).

The following books go into more detail about some of the topics that I have described here:

Carl T. Bergstrom and Jevin D. West, *Calling Bullshit: The Art of Scepticism in a Data-Driven World* (Penguin Random House, 2020)

Ananyo Bhattacharya, *The Man from the Future: The Visionary Life of John von Neumann* (Allen Lane, 2021)

I. J. Good, *Good Thinking: The Foundations of Probability and Its Applications* (Dover, 2009)

Numbercrunch

Tim Jackson, *Inside Intel: Andy Grove and the Rise of the World's Most Powerful Chip Company* (HarperCollins, 1997)

Roger Lowenstein, *When Genius Failed: The Rise and Fall of Long-Term Capital Management* (Random House, 2000)

William Poundstone, *Fortune's Formula: The Untold Story of the Scientific Betting System That Beat the Casinos* (Hill & Wang, 2005)

Simon Singh, *The Code Book: The Secret History of Codes and Code-breaking* (Fouth Estate, 2002)

Jimmy Soni and Rob Goodman, *A Mind at Play: How Claude Shannon Invented the Information Age* (Amberley, 2018)

David Spiegelhalter and Anthony Masters, *Covid by Numbers: Making Sense of the Pandemic with Data* (Pelican, 2021)

David Sumpter, *The Ten Equations That Rule the World: And How You Can Use Them Too* (Allen Lane, 2020)

Edward O. Thorp, *A Man for All Markets: Beating the Odds, from Las Vegas to Wall Street* (Oneworld, 2017)

Gregory Zuckerman, *The Man Who Solved the Market: How Jim Simons Launched the Quant Revolution* (Portfolio, 2019)

Index

Index

Index

Index

SOURCES

Page vii   epigraph from: Stanley P. Gudder, *A Mathematical Journey* (McGraw Hill, 1976)

Page 50   'World record football transfer fee'
data sourced from: https://en.wikipedia.org/wiki/
List_of_most_expensive_association_football_transfers

Page 59   'World record football transfer fee (log scale)'
data sourced from: https://en.wikipedia.org/wiki/
List_of_most_expensive_association_football_transfers

Page 64   'Daily new confirmed COVID-19 deaths'
data sourced from: https://coronavirus.data.gov.uk

Page 65   'Dow Jones value (linear scale)'
data sourced from: https://www.macrotrends.net/1319/
dow-jones-100-year-historical-chart

Page 66   'Dow Jones value (log scale)'
data sourced from: https://www.macrotrends.net/1319
dow-jones-100-year-historical-chart

Page 70   'Moore's Law: The number of transistors per microprocessor'
data sourced from: https://ourworldindata.org/grapher/
transistors-per-microprocessor

Page 82   'Phase portrait of England hospitalisations'
data sourced from: https://coronavirus.data.gov.uk/

Page 109 'Aston Villa goal probabilities'
data sourced from: https://understat.com/match/14466

Page 109 'Liverpool goal probabilities'
data sourced from: https://understat.com/match/14466

Page 114 'UK household disposable income 2020'
data sourced from: https://www.ons.gov.uk/peoplepopulationandcom-
munity/personalandhouseholdfinances/incomeandwealth/bulletins/
householddisposableincomeandinequality/financialyear2020

Page 120 Figures SPM.3 from IPCC, 2012: Summary for
Policymakers. In: *Managing the Risks of Extreme Events and Disasters
to Advance Climate Change Adaptation* [Field, C.B., V. Barros, T.F.
Stocker, D. Qin, D.J. Dokken, K.L. Ebi, M.D. Mastrandrea, K.J.
Mach, G.-K. Plattner, S.K. Allen, M. Tignor, and P.M. Midgley
(eds.)]. A Special Report of Working Groups I and II of the
Intergovernmental Panel on Climate Change. Cambridge University
Press, Cambridge, UK, and New York, NY, USA, pp. 1-19

Page 127 'Green Jelly Beans Linked to Cancer' created by xkcd, and
sourced and reproduced from https://xkcd.com/882/

Page 146 'North West COVID patients 2020 (log scale)
data sourced from: https://coronavirus.data.gov.uk

Page 148 'North West COVID patients 2020 (log scale
data sourced from: https://coronavirus.data.gov.uk

Page 196 'Smartphone Penetration of the Smartphone Market'
data sourced from: https://www.comscore.com/Insights/Blog/
US-Smartphone-Penetration-Surpassed-80-Percent-in-2016

Page 196 'Percentage of Users by Web Browser'
data sourced from: https://stats.areppim.com/stats/
stats_webbrowserxtime_eu.htm

Page 228 'Sterling vs dollar exchange rate 2022'
data sourced from: https://www.google.com/finance/quote/GBP-USD

Page 235 Drawing of the interconnectivity of chessboard squares'
data sourced from: https://en.wikipedia.org/wiki/Knight%27s_tour